CONFERENCE
ON
LANDING ASSAULTS

24 MAY—23 JUNE, 1943

U.S. ASSAULT TRAINING CENTER
EUROPEAN THEATER OF OPERATIONS
UNITED STATES ARMY

By Authority of
CO Assault Trng Cen
ETOUSA (Prov)
Initials: PWT
Date: 1 July 1943
Copy No: _____

IMPORTANT

This record is classified as "SECRET" (equals British "MOST SECRET") This classification covers the Record, as such, and all papers and documents contained herein.

It should not be dismantled from its present form and no sections thereof should be removed without the classification of "SECRET" being clearly designated.

TABLE OF CONTENTS
VOLUME TWO
PHASE III

FIELD EXERCISES

Exercise No. 1.

Exercise No. 2.

Exercise No. 3.

PHASE IV

Adaptation of FM 31-5

Proposed Training Circular

CONCLUDING REMARKS

Concluding Remarks   Brig Gen Daniel Noce

                                Col P.W. Thompson

The Chairman's   Lt Col L.P. Chase
Summation

APPENDIX

List of Conferees
Note to Conferees
Invitation to Speakers
Organization
General Information
Schedules
Committees - Phase III
Directive - Phase III
Committees - Phase IV

PHASE III

EXERCISE 1

ASSAULT TRAINING CENTER
CONFERENCE

HQ ETOUSA

1 July 1943

FIELD EXERCISES

The Field Exercises comprising Phase III of the Conference, as prepared by the three committees appointed by the Conference Chairman, are presented herewith.

These committees spent the period from 8 June to 13 June in the Woolacombe Area preparing their exercises.

ASSAULT TRAINING CENTER
CONFERENCE

HQ ETOUSA

1 July 1943

RECOMMENDATION BY COMMITTEES

The following recommendations were made jointly by the three committees assigned to develop field exercises for a landing assault.

RECOMMENDATION 1: Special Assault Platoons should be developed.

RECOMMENDATION 2: The scale for different operations will vary, but for purposes of procurement and planning, it is our recommendation that a division landing on a 4 battalion front should be furnished support craft at least on the following scale:

        8 LCG
        8 LCS (L)
        4 LCF
        8 LCT (R)

RECOMMENDATION 3: Ranger training should be started at once.

All three committees used tanks in the first wave with the Assault Platoons.

ASSAULT TRAINING CENTER
CONFERENCE
HQ ETOUSA

21 June 1943
KM

Exercise No. 1

General Situation

1. The BARNSTAPLE BAY area is assumed to be part of the coast of FRANCE. The coast is assumed to be as shown on the map between MORTE POINT and HARTLAND POINT (both inclusive). It is assumed to run due north from ROCKHAM BAY (on the north side of MORTE POINT), and due south from HARTLAND POINT. All is land east of this assumed coastline.

2. The enemy holds the entire coast, in both directions from the BARNSTAPLE BAY area, in considerable strength, divisions holding sectors measuring about 25 to 35 miles in width. All sectors are topographically similar to the BARNSTAPLE BAY sector, and are fortified on about the same scale.

3. A panzer division is located in the EXETER area. This division can close in the BARNSTAPLE BAY area within 24 hours after the beginning of an assault. Another panzer division can be expected to close in the area within 48 hours of an assault.

4. The I Corps (reinforced) has been designated to invade FRANCE near BARNSTAPLE BAY. The I Corps is at present located in the vicinity of embarkation points approximately 80 miles by water from the coast of FRANCE.

First Special Situation

1. On   the CG 1st Infantry Division was called to the Headquarters I Corps, where he received the following information:

   a. The I Corps with 1st and 2d Infantry Divisions abreast as Assault Divisions (1st Inf Div on the left) will invade FRANCE near BARNSTAPLE. (See operations map).

   b. D-Day will be announced later.

   c. H-Hour - 3 hours prior to nautical twilight.

   d. NAVAL SUPPORT   The Navy is prepared to support the landing by the following means:

   (1) Furnish naval gunfire support.

   (2) Furnish support craft of the following types:

      (a) LCS(S)

      (b) LCS(M)

      (c) LCS(L)

      (d) LCT(R)

      (e) LCF(L)

      (f) LCG

(3) Trained Naval shore fire control parties each consisting of 1 Army Officer, 2 Naval Officers, and 4 Privates with necessary communication equipment will be furnished on the basis of one team for each battalion landing team.

(4) One Army gunfire support liaison officer will be aboard each firing ship.

e. AIR SUPPORT All indications point to definite air superiority for the landing forces. However, it is known that the enemy has available a large enemy fighter force and at least 150 bombers prepared to oppose both the passage of the convoy and the landing proper. Despite our air superiority isolated air attacks can be expected.

The Air Force is prepared to support the landing by the following means:

(1) Preparatory bombardment from D-7.

(2) Provide fighter cover and escort except during hours of darkness in amounts required from time convoy leaves embarkation point.

(3) Night area bombing missions.

(4) Day low level bombing missions.

(5) Day high level bombing missions.

(6) Fighter bomber missions.

(7) Dive bomber missions.

(8) Photo reconnaissance missions.

(9) Smoke laying missions.

(10) Cannon fighter straffing.

(11) Transport of such airborne units as may be required.

(12) Provide antisubmarine patrol.

The 1st Infantry Division will be furnished four (4) trained Air Support Parties.

Air Support Control will be operated at Headquarters I Corps.

f. 1st Infantry Division reinforced will land in its zone of action on D-DAY at H-HOUR and seize and hold the line of the RIVER YEO (BARNSTAPLE exclusive) as a beach head line. See operations map. Maintain contact with 2d Div on right (south) and II Corps on left (North). CG 1st Inf Div responsible for establishment and operation of beach maintenance area until relieved by orders of CG, I Corps.

g. Composition 1st Infantry Division, reinforced.

See Annex #1, attached.

h. Scales of MT personnel and equipment

(1) Motor Transport: Motor transport will be limited to assault scale and to following types of vehicles:

        1/4 ton Jeeps (Land and amphibious)
        3/4 ton Command and Reconnaissance
        3/4 ton Weapons Carriers

2.1/2 ton Trucks (Land and Amphibious)
1/4 ton Trailers.
1 ton Trailers.
Such special vehicles as required by Beach Maintenance Group.

(2) <u>Personnel</u>: Personnel will be figures on the elimination from T/O strength of the following:

Bands
Basics
Chauffeurs for motor vehicles remaining at home stations
Motorcyclists
Reduction of rifle squads to 10 men.

Final T/O strength after above eliminations will be reduced by 3% for figuring boat space required.

In no case will tactical units such as platoons and companies be eliminated.

(3) <u>Equipment</u> -(<u>a</u>) Organizational equipment will be reduced to the minimum required for combat.

(<u>b</u>) Personal equipment will be reduced to absolute minimum for combat over a period of three (3) days.

(<u>c</u>) In determining required equipment, the following factors will be considered:

Assault troops are physically fit and tough and can live on emergency rations without loss of combat effectiveness for a period of three (3) days.

Assault troops must be prepared to fight immediately they leave landing craft.

Means must be provided for passage of underwater and beach obstacles, and minefields.

Means must be provided for attack of concrete gun emplacements, and destruction of antitank obstacles.

The enemy may use gas.

FIRST EQUIPMENT

1. Tactical plan prepared by CG 1st Infantry Division (show scheme of maneuver on operations map).

2. Number and type of motor transport to be embarked for landing on D-day.

3. Administrative plan of CG 1st Infantry Division to include means of providing for the period D to D / 3 of:

<u>a</u>. Rations.

<u>b</u>. Water

<u>c</u>. Ammunition

<u>d</u>. Care and evacuation of wounded

4. Requests made by CG 1st Infantry Division relative to:

   a. Naval gunfire support desired.

   b. Support craft desired.

   c. Air support desired.

   d. Type and number of ships and craft required to transport division and attached troops.

   e. Special equipment (not included in Table of Basic Allowances) required for the division and attached troops.

5. Show by sketch plan for beach maintenance area to include:

   a. Beach exits.

   b. Location of Dumps.

   c. Medical establishment.

   d. Transportation Park.

   e. Communication plan.

   f. Assembly areas for "follow up" troops.

SECOND REQUIREMENT

1. Based on solution of FIRST REQUIREMENT, give the tactical plan prepared by the CO, CT 1 (show scheme of maneuver on operations map).

THIRD REQUIREMENT

1. Based on solution of SECOND REQUIREMENT give the tactical plan prepared by the CO 2d Bn 1st Infantry Regiment (show scheme of maneuver on operations map).

2. Show by diagram the boat formation used by this battalion during approach to shore. Include loading of each boat.

FOURTH REQUIREMENT

1. Based on solution of THIRD REQUIREMENT state pars 2 and 3 of Field Order issued by CO E Company 2d Bn, 1st Infantry.

FIFTH REQUIREMENT

1. Based on solution of FOURTH REQUIREMENT state pars 2 and 3 of Orders issued by CO 2d Platoon E Company 2d Bn 1st Infantry.

SIXTH REQUIREMENT

1. Based on solution of FIFTH REQUIREMENT state pars 2 and 3 of Orders issued by the Squad leader of 1st Squad 2d Platoon E Company 2d Bn 1st Infantry.

ANNEX I TO GENERAL & SPECIAL SITUATION

1. Division Headquarters and Special Troops (less detachment to 3 Regimental Landing Teams).

Unit	T/O	Off	W.O.	E.M.	Aggregate
Division Headquarters	7-1	44	9	116	169
Headquarters Company	7-2	7	3	134	144
Military Police Platoon (less Det)	19-7	3	–	41	44
Reconnaissance Troop (less Det)	2-67	4	–	59	63
Signal Company (less Det)	11-7	10	1	275	286
Engineer Battalion (less Det)	5-15	13	2	156	171
Medical Battalion (less Det)	8-15	20	–	164	184
Quartermaster Company	10-17	10	–	196	206
Ordnance Company	9-8	9	1	145	155
Division Artillery Hq	6-10	8	1	28	37
Division Artillery Hq Bty	6-10-1	2	1	103	106
Medium F.A. Battalion	6-35	28	2	611	641
TOTALS		158	20	2028	2206

2. Attached Units (less detachments with Regimental Landing Teams)

Unit	T/O	Off	W.O.	E.M.	Aggregate
Tank Destroyer Battalion	18-25	35	–	860	895
* Engineer Regt. (less Det)	5-171	15	2	172	189
Tank Battalion (Light)(less Det)	17-75	18	2	206	226
– AA Battalion (37 mm)	4-25	32	2	810	844
– AA Battalion (37 mm)	4-25	32	2	810	844
101st Airborne Division	71	506	29	7970	8505
501st Parachute Regiment	7-31	140	5	1884	2029
Three (3) Ranger Bns		90	0	1500	1590
TOTALS		868	42	14,272	15,122

* Engineer Regiment with three (3) Bns

3. Regimental Landing Team (each)

Unit	T/O	Off	W.O.	E.M.	Aggregate
Infantry Regiment (less Band)	7-11	143	5	3295	3443
Artillery Battalion (light)	6-25	29	2	562	593
Collecting Company Med Bn	8-17	5	–	102	107
Engineer Company	5-17	5	–	194	199
** Tank Company (light) Reinf	17-17	6	–	129	135
Platoon Rcn Troop	2-67	1	–	45	46
Detach Sig Co				12	12
Shore Fire Contact party		6	–	18	24
Traffic Squad MP Plat	19-7			12	12
Shore Party Engineer Bn	5-171	17	–	590	607
Signal Detachment		1	–	60	61
*** Attached medical		3	–	31	34
**** Naval Det		6	–	129	135
TOTALS		222	7	5179	5408

** Attached from Hq 19 EM, Ser 2 EM, Med. Det. 3 EM
*** To permit 1 off and 10 EM per Engineer Company with Battalion Landing Team
**** Naval Detachment made up of 1 Off (Naval), Beachmaster, 1 Off (Naval) Medical, and 43 EM (Naval) for each.

4. Each Regimental Landing Team is embarked with two (2) Battalion Landing Teams in AP's and the remainder in LCI (S) and LCT's or LST's as follows:

Battalion Landing Team (each)

	T/O	Off	W.O	E.M.	Aggregate
Infantry Bn	7-15	34	-	920	954
Platoon Canon Co	7-14	1	-	33	34
Platoon Anti-tank Co	7-117	1	-	34	35
Mine Squad AT Co	7-17			8	8
Trans Sec Ser Co	7-13	1	-	11	12
Arty Bty (light)	6-27	4	-	110	114
Shore Fire Control Party		2	-	6	8
Platoon Engineer Co	5-17	1	-	53	54
Platoon Tank Co (L) Reinf		1	-	28	29
Det Med Bn				6	6
Shore Party					
Engineer Co	5-17	5	-	194	199
Signal Det				18	18
Med Det		1		10	11
Naval Det		2		43	45
TOTALS		53		1474	1527

Total Divisions and Attached Units

	Off	W.O	E.M.	Aggregate
Division less 3 Regt Lts	158	20	2028	2206
Attached troops less Dets	868	42	14212	15122
3 Regimental C.T's	666	21	15537	16224
GRAND TOTALS	1692	83	31777	33552

HEADQUARTERS
ASSAULT TRAINING CENTER
ETOUSA (PROV)

Solution to Par 1, First Requirement

2. This Division reinforced (see Annex #1) will land at beaches between the river TAW estuary and BULL POINT, quickly capture and (improve) beach exits near WOOLACOMBE and CROYDE BAY, seize and hold the line of the RIVER YEO (BARNSTAPLE exclusive) and establish and maintain a beachhead for further operations.

    Landing Beaches: See Operations Map
    Formation   : See Operations Map
    Boundaries   : See Operations Map
    Time of landing: D-day (to be announced later); H-hour, 3 hours before nautical twilight.

3. **a.** CT 2 (see Annex #1 attached) will land at Beaches "A" and "B" immediately capture the beach exit at CROYDE BAY and in accordance with scheme of maneuver indicated on Operations Map, seize the objectives within the zone of action of the CT. It will protect the right flank of the Division and maintain contact with the Division on the right. CHIVENOR AIRFIELD will be captured with utmost speed, and Division Hq notified immediately on capture of same. Advance beyond first objective will be made only on orders from Division.

    b. CT 1, with 3d Ranger Bn attached (see Annex #1) will land at Beaches "C" and "D"; immediately capture the beach exit near WOOLACOMBE and, in accordance with the scheme of maneuver indicated on Operations Map, seize the objectives within the zone of action of the CT. It will protect the left flank of the Division and maintain contact with the unit on the left. Advance beyond first objective will be made only on orders from Division.

    c. The CT 3 (see Annex #1) in floating reserve will be prepared to land on orders in the zone of action of either assault regiment. For this purpose it will establish and maintain liaison with each assault regiment.

    d. The 1st Ranger Bn will land along the coast near BAGGY POINT at H-1 hour and reduce coast artillery guns located near CROYDE HOE. (For Bomber support see Air Force Annex). The Bn will then assist the 2nd CT in capturing the beach exit at CROYDE BAY by flanking action to the north. Attached to CT 2 after capture of beach exit at CROYDE BAY.

    e. The 2nd Ranger Bn. will land at Beach "E" at H-1 hour, capture the hostile coast artillery guns located near MORTE POINT and then assist CT 1 in capturing the high ground north of WOOLACOMBE by flanking action to the south. (For Bomber support see Air Force Annex). The 2nd Ranger Bn will establish contact with the 3d Ranger Bn which will land at Beach "C" at H hour, and thereafter is attached to following CT 1.

    f. Division Artillery (less CT attachments) will land at Beach "D" immediately following CT 1 and support the attack with special attention to the support of CT 1. The CG Division Artillery will make plans for bringing all supporting artillery under his control with least practicable delay after landing.

    g. Anti-aircraft Artillery (less CT attachments) will land at Beach "B" immediately following CT 2. It will proceed without delay to

CHIVENOR AIRFIELD and provide anti-aircraft protection for the airfield.

Anti-aircraft Bn attached to combat teams will provide anti-aircraft protection for the beaches in zones of action of respective combat teams. This Battalion will revert to control of CO 1st AA Battalion after landing. CO 1st AA Battalion will prepare anti-aircraft defense plan for protection of the beaches and beach maintenance area. He will report to Division CP for instructions immediately same is established in shore.

h. 1st Reconnaissance Troop is attached to CT 1 for landing. On landing it will assemble vicinity of WOOLACOMBE and perform reconnaissance mission as ordered in G-2 plan (omitted).

i. Tank Destroyer Battalion is attached to CT 3 for landing. Initially it will be in floating reserve with CT 3. After landing it will remain attached to CT 3 until receipt of further orders.

j. 1st Tank Battalion (light) (less CT attachments) will land following Division Artillery on Beach "D". CO 1st Tank Battalion (Light) will coordinate initial employment of Tank Companies attached to Combat Teams with respective CT Commanders. Upon arrival of CP 1st Tank Battalion (Light) in shore all Tank elements will revert to control of the CO 1st Tank Battalion.

k. 1st Engineer Bn (less CT attachments) is attached to CT 1 for landing. Plan of employment of units of 1st Engineer Battalion attached to CT's will be coordinated by CO 1st Engineer Bn with respective CT Commanders. Units of 1st Engineer Bn attached to CT's will revert to control of CO 1st Engineer Bn when the latter has established his CP ashore.

l. The 101st A/B Division will secure the high ground extending from R-J 679 to HORE DOWN GATE, incl, and block all movement of hostile reserves from the east across the BARNSTAPLE - BITTADON - ILFRACOMBE ROAD. It will establish contact with CT 1 and assist that CT in the capture of the high ground 1000 yards south of WILLINGCOTT.

Parachute elements of the division will initiate landings at H-hour, followed by glider elements at H plus 3 hours.

All elements remain east and north of the SOUTHERN RAILWAY LINE until H plus 4 hours.

Dropping Zone - along line of BARNSTAPLE - BITTADON - ILFRACOMBE ROAD.

m. The 501st Parachute Regiment, landing at H-hour will capture and destroy the 105 mm howitzer battery reported $1\frac{1}{4}$ miles west of HILL 640, and will capture the BUTTERCOMBE strong point.

All elements remain east and north of the SOUTHERN RAILWAY LINE until H plus 4 hours.

Dropping Zone - vicinity HILL 640.

n. Service and Shore Party Units (see Administrative Plan omitted).

o. Landing Tables will be prepared by respective CT Commanders and submitted to Division Headquarters for coordination by G-4 and Principal Beach Master (CO Engr Regt Shore Party) by

p. Landing Tables of CT's will provide for landing in following priorities:

1. Combat Troops.
2. Shore Party Dets (in this connection all are reminded that elements of Shore Party Dets must be landed with Combat Troops).
3. Fighting Vehicles.
4. Supplies

Landing Tables of Combat Team must provide for having ashore by H plus 6 of all Combat Troops including attached AA Artillery of their respective Combat Teams. In addition the following units will be landed by H plus 6:

        Advance Det Division Headquarters
        Hq Division Artillery
        Division Artillery (less CT attachments)
        1st Tank Bn (Light) (less CT attachments)
        1st Engineer Bn (less CT attachments)
        All AA Artillery
        1st Reconnaissance Troop

<u>See Operations Maps.</u>

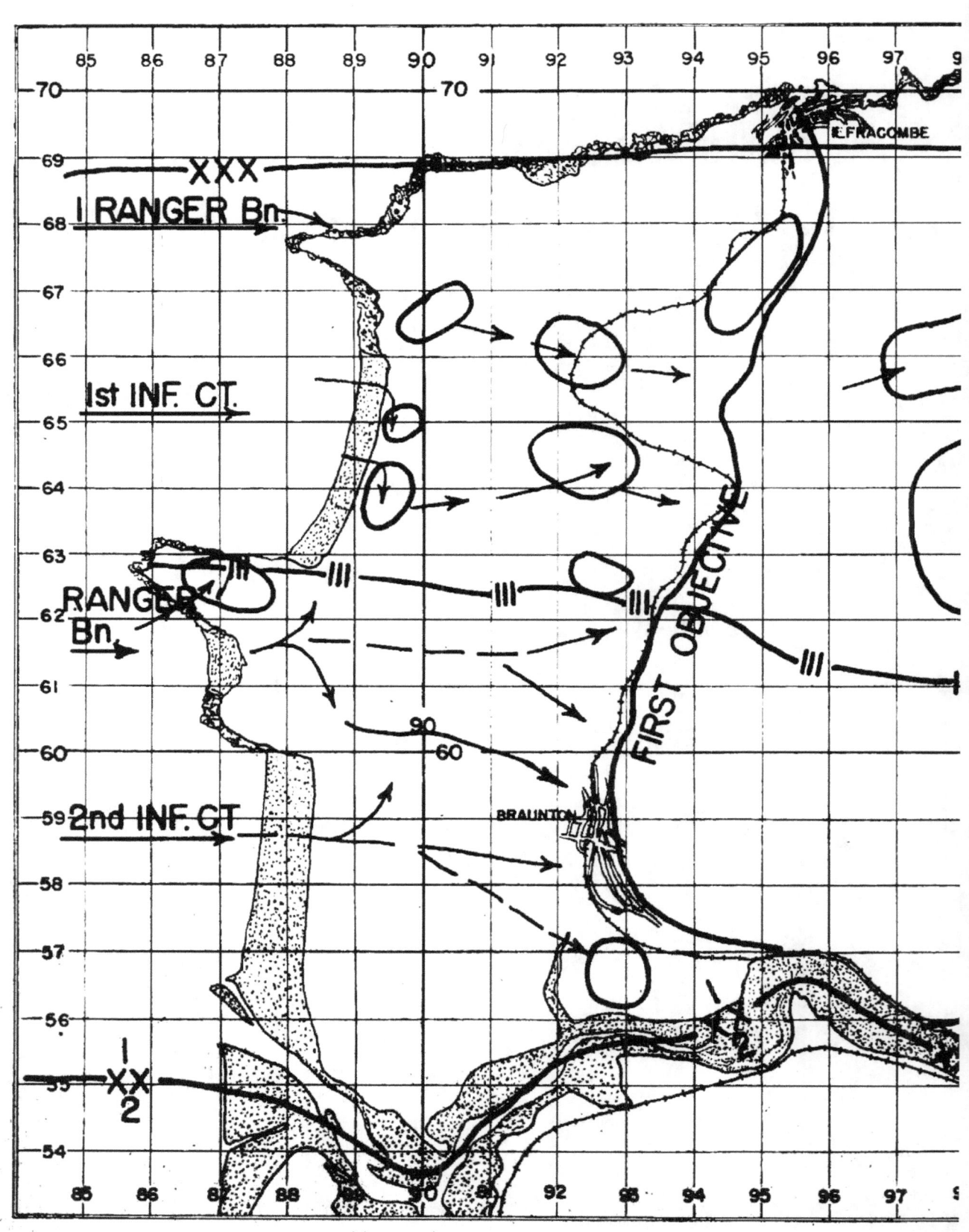

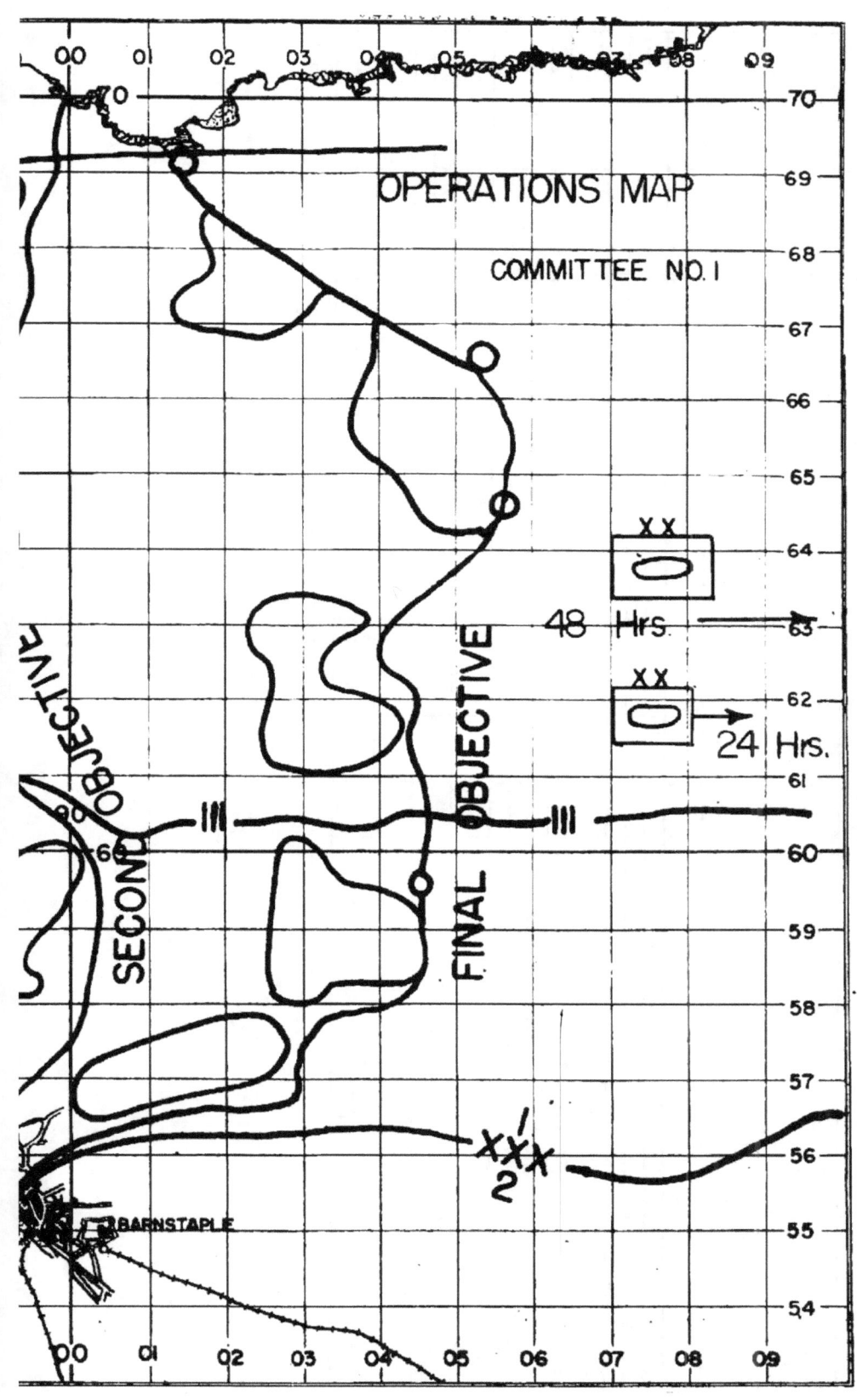

FIRST REQUIREMENT PAR #2

Bn Landing Group	Personnel		Vehicles												Towed Weapons			Self propelled Weapons			
	Officers	Enlisted Men	¼ t Tk.	¼ t Trl.	¾ t Tk W.C.	¾ t Tk C & R	1 t Trl.	Ambulance	Half Tractor	Bulldozer	Lt Tk.	2½ t Tk.	Water Trl.	TOTAL	75 mm How	37 mm Gun	TOTAL	AA (AW)	75 mm	105 mm	TOTAL
Hq & Hq Co	9	130	19	5																	
3 Rifle Cos	18	576	6	6																	
H W Co	5	178	19	14												4	4				
Bn Med Sect	2	36	2	2	1																
Trans Sect																					
Serv Co	1	11	1		1		6														
Btry FA *	6	116	5		7	1	3								4	2	6				
Det Med Bn		6						1													
Tk Plat	2	69	1						1		5										
Mine Sq A T Co	1	28	1																		
Eng Plat (combat Eng)		8			3																
A T Plat (attchd)	1	53	1	1	4										1	4	4				
Plat Cannon Co	1	34	2				3												3		3
Eng Co (Sh Btry)**	1	33	2							2		2									
			59	27	22	1	12	1	1	2	5	2									

* Includes shore Fire Control Pty.
** Does not include Naval Det.

SOLUTION TO FIRST REQUIREMENT PAR #2

Regt'l landing Group (less 3 Bns)	Personnel		Vehicles												Towed Weapons			Self propelled Weapons			
	Officers	Enlisted Men	1/4 t Trk.	1/4 t Trl.	3/4 t Trk W.C.	3/4 t Trk C & R	1 t Trl.	Ambulance	Half Tractor	Bulldozer	Lt. Tk.	2½ t Tk.	Water Trl.	TOTAL	75 mm How	37 mm Gun	TOTAL	AA (AW)	75 mm	105 mm	TOTAL
Hq & Hq Co	17	118	11	3	1	1	1					2			1						
Service Co (-)	10	86	3	4	11		5					8									
Cannon Co (-)	2	26	4	4	4	1						4							6	6	
A T Co	4	35	3	3	12	1	2					2				12					
Med Det (-)	4	18	4	3								1									
Hq & Hq Bty F A Bn	14	159	5	2	6	2	4	1	3			6				6				4	
Service Bty F A Bn	5	73	4			2	1					9									
Medium F A Btry	5	118	4		5	1	1		5	1		4									
Eng Co (-)	2	35	1		3							2									
Tank Co (-)	3	45	1								1	2				1					
Det Sig Co		12				3	1					1	1								
Eng Bn (-) (5P)	3	15	1	1		1	1		2	1		1									
Det MP Plat		12	2																		
Coll Co	5	84	1	1		3	3	12				4									
Ren Tr Plat	1	45	10	4							1					1					
			54	20	42	15	18	13	10	1	1	45	1		2	19			6	10	

-11-

SOLUTION TO 1st REQUIREMENT - WRAP. 3
ADMINISTRATIVE ORDER

ADM Order # 1 to accompany Field Order # 1.

Maps See Operations Map F.O. #1.
References: FTP 167
     FM 100-10
     FM 31-5
     FM 8-25

1. General.

 a. Restrictions imposed by the employment of a limited number of ships, craft and planes and the high ratio of less preclude a possibility that the ordinary logistical means will be available during the initial stages of this operation. Landing team, combat team and division commanders are responsible for the logistical support of their respective units until higher echelon supply and evacuation facilities are available.

 b. Since all supplies which will support this attack must be transported from a distance by air or over water and be available to troops immediately upon landing, either from air or sea, adequate ammunition must be loaded at ports of embarkation to be available for issue prior to landing; vehicles must be loaded with organic loads (including ammunition), and planes, ships and craft employed must be loaded to conform with the tactical scheme. Because a difference exists with respect to the methods of supply the plan for this operation is treated in two parts:

  (1) Airborne troops
  (2) Amphibious borne troops.

2. SUPPLY (Amphibious Borne Troops)

 a. Embarkation Procedure.

  (1) Reinforced regimental combat teams in assault plus Ranger Bns, will be combat unit loaded. Other divisional elements and attack troops will be organizational unit loaded.

  (2) Supplies for all but Ranger Bns will be loaded in such a manner as to provide self sufficiency for a period of 10 days. Ranger Bns will load supplies for 3 days. Loading will be planned and accomplished with deliberate care, contributing in every possible manner to progressive support of the tactical plan. Embarkation forms will be scrupulously followed in the progress of loading. Embarkation forms will be submitted as follows by noon - Jun.

   a. Unit Personnel and Tonnage.

   b. Loading and Storage Planes; Consolidated Tonnage Table.

   c. Organization of embarkation troops and transport divisions.

  (3) Supplies to be carried.

  (a) Ammunition.

   (1) Quantity - 5 units of fire.
   (2) Distribution - 5 units of fire within each reinf regimental combat team as follows:

       (a) 3 units of fire for each battalion landing team.
       (b) 2 units of fire for each RCT or AK's and LSTs.
       (c) 5 units of fire with remaining divisional elements and attached troops.

   (b) Rations.

     (1) 5 landing rations for force will not be carried.
     (2) Emergency rations for embarked troops loaded w/each units 6 days K ration.
       4 " C "
       1 " D "

   (c) Gasoline (both initial and resupply by Navy)

     (1) Gasoline to be provided in two types:

       (a) Unloaded (suitable for field ranges)
       (b) Standard motor fuel (suitable for wheeled vehicle or tank operation)

     (2) Approximately 10 days gasoline carried within each RCT packed in 5 gal expeditionary cans and in drums in the ratio of 3 to 4.

     (3) Gasoline for units organizational unit loaded will be packed in drums.

     (4) 5 gal expeditionary containers (filled) will be in hands of units embarked.

     (5) 50 empty drums (new) per AK

     (6) Bulk resupply by Navy from AOG (gas tankers)

     (7) Gasoline in expeditionary cans and drums will be stencilled as follows:

       (a) Unleaded gasoline with the word "white" in blue paint.
       (b) Motor fuel with the word "motor" in red paint.

   (d) Water.

     (1) To be provided by naval vessels until adequate shore sources are developed.

     (2) 1 days water at 2 gallons per man/day packed in 5 gallon expeditionary cans for embarked troops combat loaded.

     (3) Storage facilities for 2 days at 2 gallons /man/day to be embarked w/ea transport division (supplied by troops).

     (4) Distillation and water purification equipment will not be provided by this division.

     (5) Navy to provide water tanks and pumps for bulk movement to shore. Each landing team shore pty will be equipped with pumping equipment to pump water into storage tanks.

   (e) Other Supplies.

     (1) No combat replacement supplies will be loaded for RCT's.

     (2) 30 days replacement supplies of all classes embarked within division excepting special amphibious equipment

(division reserve).

(f) Motor vehicles.

(1) Will be embarked in manner best calculated to support tactical debarkation.

(2) Each vehicle will be stocked prior to embarkation with its combat load, gasoline tank full, 10 gallons reserve gasoline in expeditionary containers, 1 gallon reserve oil and 1.1/3 K ration for driver and assistant.

(3) Vehicles will be carefully checked prior to embarkation for fuel leaks and electrical short circuits. Each vehicle will be equipped with ground wire for grounding when embarked. Fan belts will be disconnected prior to embarkation.

(4) Battery units, generators, wheel bearing, springs and shackles, and universal joints will be covered with base grease prior to departure from the home station.

(5) Motors of vehicles will be started at least one hundred (100) yards from the beach and brakes firmly set.

(g) Marking equipment and supplies.

(1) In addition to marking prescribed by the Port of Embarkation, all boxes or crated equipment and supplies in hands of units will be marked as follows only:

(a) Box color and organizational identification as approved by this headquarters.

(b) Pieces to which access is required en route will be painted with a 3" yellow disc on top and sides.

(c) Pieces to which access is not required en route will be painted with a 3" white disc on top and sides.

(d) The unit personnel and tonnage table line number under which the contents of the boxes fall, painted in red on the yellow or white disc as the case may be.

b. Supply Procedure.

(1) General

(a) Organizational unit loaded vessels will be unloaded when and as indicated by this headquarters.

(b) Bn landings teams and senior units will establish beach dumps and evacuation facilities in accordance with the tactical situation.

(2) Control and responsibility.

(a) The senior commander ashore in each subordinate zone of action during the landing attack will coordinate and be responsible for the functioning of the supply and evacuation systems within that zone.

(b) Rear boundaries will be prescribed as the operation develops.

(c) System of supply.

(1) Ammunition

(a) Landed as directed by appropriate

commander.

(~~~) ~~All communications~~ sent ashore will be considered expended until reported on hand by an operating unit.

      (c) Ammunition landed will be moved into selected dumps by shore pty and issued by supply elements of senior unit ashore and operating in that zone.

    (2) Rations.

      (a) Assault troops carry 1.1/3 K ration.

      (b) Unit kitch trucks stocked with 10 2/3 K and 10 rations per entire unit.

      (c) Remaining rations will be landed in priority as follows:

        4 K rations
        3 C   "
        1 D   "

      (d) Rations will be consumed as follows:

D-Day K ration
D / 1 day

    Breakfast K ration (breakfast unit)
    Dinner   K ration (dinner unit)
    Supper   1/3 C ration plus 1/9 D ration

Successive days

    Breakfast 1/3 C ration
    Dinner   K ration unit (1/3 K ration)
    Supper   1/3 C ration plus 1/9 D ration

      (e) Ration cycle commence with supper.

    (3) Gasolone and oil

      (a) Landed initially in expeditionary containers. Resupply in 50 gallon drums.

      (b) Vehicles carry 10 gallons reserve of gasoline and 1 gallon reserve oil at all times.

      (c) Gasoline in 5 gallon expeditionary containers will be maintained at supply points for all classes of supply. Vehicles replenish by exchange of empty for full containers.

    (4) Water

      (a) Assault troops carry 2 filled canteens.

      (b) Initially transported ashore in 5 gallon expeditionary containers.

      (c) Bulk water moved ashore in tanks. Issued to troops in 5 gallon expeditionary cans by exchange of empty for full containers.

      (d) All local water used will be chlorinated. Units down to and including signals will be provided with emergency chlorinated material.

  d. Evacuation

    (1) Casualties (Personnel)

(a) Until facilities are established ashore naval medical sections of shore parties will provide first aid and evacuate all personnel casualties to ships and craft designated.

  (2) Burial

    (a) By division
    (b) No cemeteries are designated at this time.
    (c) Enemy dead will be interred in the same manner as our own.

 e. Motor maintenance.

  (1) Units perform all maintainence possible. Regiments and separate units will carry extra supplies of spare parts to assist in maintenance for the period herein before setforth.

 f. Ordnance Maintenance.

  (1) Maintenance of ordnance material will be carried out with such tools and spare parts as are available and by canabalization when necessary.

 g. Prisoners of war.

  (1) The senior commander ashore will be responsible for evacuation of prisoners to ships.

 h. Traffic.

  (1) By subordinate units within their assigned areas.
  (2) Traffic control in beach area by shore party.
  (3) Road priority

    (a) Combat vehicles.
    (b) Wire laying vehicles and ambulance.
    (c) Ammunition carrying vehicles.
    (d) Staff and messenger vehicles.
    (e) Gasoline and water carrying vehicles.

 i. Miscellaneous.

 j. Salvage.

  (1) No attempt will be made to salvage except in so far as such is necessary for canabalization.

 k. Captured material.

  (1) Every attempt will be made to employ captured material where possible.

  (2) No attempt will be made to collect such material.

 l. Personnel

  (1) Stragglers

    (a) The senior commander ashore in each subordinate zone of action will establish a straggler line and collecting points.

  (2) Mail - there will be no collection of mail until D/10 Mail will be distributed whenever possible.

  (3) Strength reports.

    (a) Strength reports will be submitted upon embarkation.
    (b) Casualty reports will be submitted as required.

(4) Replacements

    (a) There will be no replacements until relief.

3. SUPPLY (Air Borne troops)

  a. General.

    (1) Elements of Air Borne Divisions and attached troops will be loaded in accordance with Adm SOP # 1 dated------

    (2) Supplies will provide self sufficiency for a period of six (6) days combat and administration.

SOLUTION TO FIRST REQUIREMENT
PAR 4a.
SCHEDULE OF FIRES
FIRE SUPPORT GROUP NO. 1
(OFFSHORE)

Target No.	Sector	Type Fire	Time From	To	100 yd squares	Mins & Density	Amm Required	Remarks
1	B	CB	H-120	H-60	5	--	as reqd.	Note 1
10	B	CB	H+200	H+206	3	--	as reqd.	Note 2
16	B	CB	H+200	H+206	3	--	as reqd.	Note 2
13	B	CB	H+200	H+206	3	--	as reqd.	Note 2
11	B	CSF			6	3x12	8 =12" 18 = 5"	
12	B	CSF			6	3x12		
14	B	CSF			6	3x12		
17	B	CSF			6			
15	B	CSF			6			
13	B	CSF			4			

FIRE SUPPORT GROUP NO. 2

Target No.	Sector	Type Fire	Time From	To	100 yd squares	Mins & Density	Amm Required	Remarks
20	A	CB	H-120	H-60	6	--	as reqd.	Note 1
29	A	CB	H+200	H+206	3	--		Note 2
30	A	CB	H+200	H+206	3	--		Note 2
45	A	CB	H+208	H+214	3	--		Note 2
46	A	CB	H+208	H+214	3	--		Note 2
44	A	CB	H+208	H+214	3	--		Note 2
40	A	CSF			6			
41	A	CSF			6			
42	A	CSF			6			
43	A	CSF			6			

Note 1: CB fires only in case coast defense btry opens fire
Note 2: Thereafter on call
Note 3: Both Support Groups be prepared to fire.
DSF and CSF on call after H plus 3 hrs 20 minutes.

    Inshore Fire Support Group No. 3 will be prepared to fire CSFs on targets 5, 7, 8 and 9, and other targets requested, on call after H plus 3 hrs 20 minutes.
    Inshore Fire Support Group No. 4 will be prepared to fire CSFs on targets 27, 28, 29, 30, 34 and 32, and other targets requested, on call after H plus 3 hrs 20 minutes.
    Support craft assigned to each Regt CT will be responsible for targets 2, 3, 4, 6, 21, 22, 23, 24, 25, 26 and 33, using point blank fire and firing only at observed flashes. Support craft will also take under fire other targets observed firing from the seaward slopes of the landing areas.

FIRE CONTROL MAP
EXERCISE #1

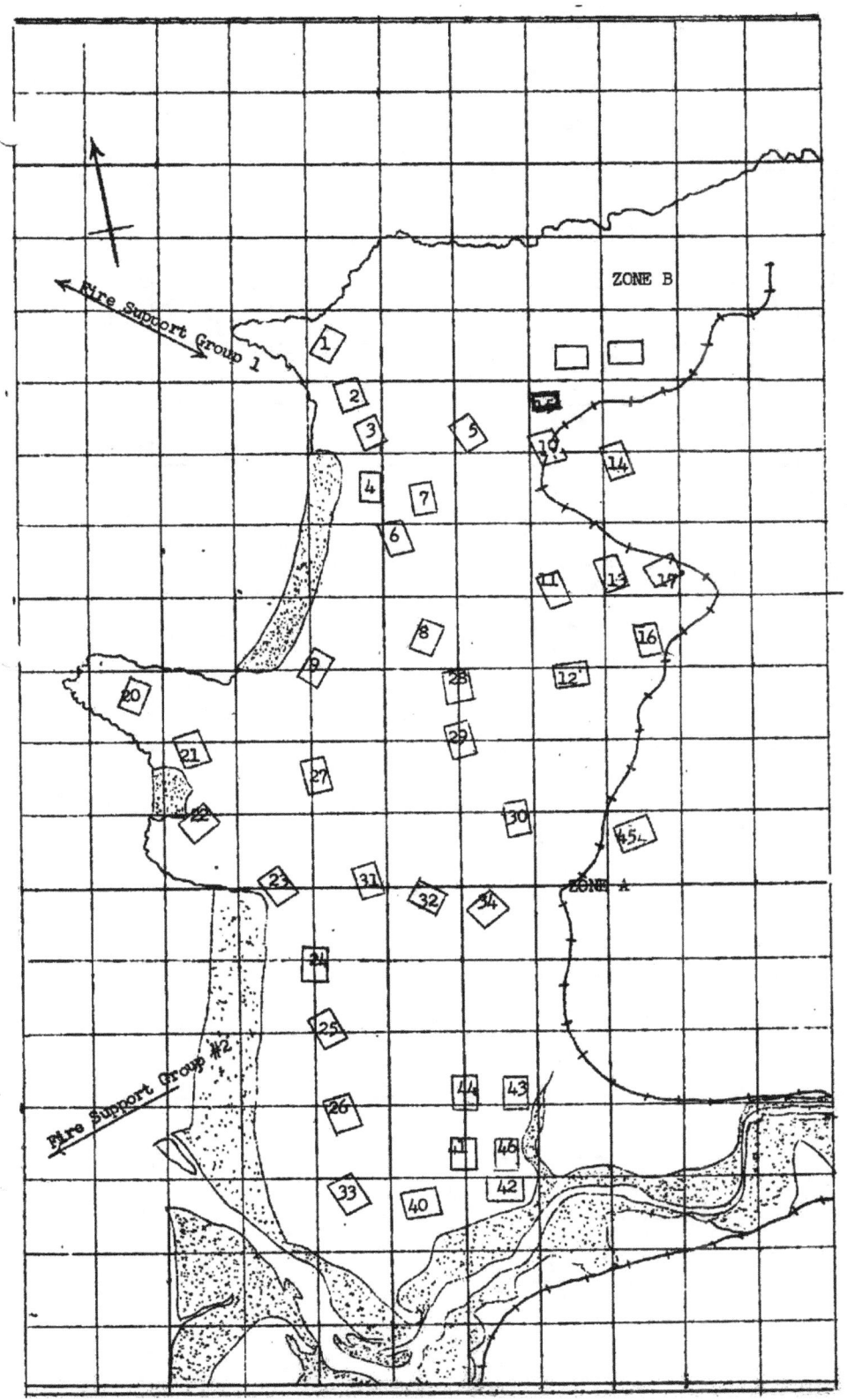

SOLUTION TO FIRST REQUIREMENT
PAR 4b. AND 4d.

LANDING SHIPS - CRAFT

ASSAULT REQUIREMENTS (2)

Type	Ranger Bn (3)	Assault Bn (4)	Reserve Bn (2)	Res Regt (1)	Div Hq-Spec & attchd troops (less det)	Total Division
APA		1				4
AKA		1				4
LCVP		(30)				120 APA
LCM (3)		(3)(8)				32 AKA / 12 APA
LCS (S)		(2)				8
LCC		(1)				4
LCR (L)	50					150
LCI (L)	3					9
LST		------ From Far Shore ------	4	12	12	32
LCT (5)		5	4	12	12	52
LCT (R)		1				4
LCG		4				16
LCF		1				4
LCS (L)		2				8
LCI (L)			4	12	12	32

Solution to Par 4c. First Requirement

The air action in support of an assault landing in general fits into two categories, <u>a.</u> Fighter cover for protection of ground and sea forces from enemy air action, and <u>b.</u> close support of ground troops. All fighter cover as in <u>a.</u> will be the responsibility of Fighter Command, while the support noted in <u>b.</u> will be a responsibility of the Air Support Command delegated to support the assault division (or divisions as the case may be). Air anti-submarine partrol will be a responsibility of Coastal Command.

Early air action will be controlled by the Air Force representative from the Hq Ship and by the Fighter Director Ship within their respective spheres of action. Movement of such control from the ship to shore installations will take place as soon as the situation permits.

Four air support parties are necessary, these would be allocated as follows:

<u>a.</u> One to each assault regiment (two)
<u>b.</u> One to the reserve regiment
<u>c.</u> One to division headquarters.

The AS party with the reserve regiment will constitute additionally a logical replacement for casualties in the AS parties of the assault regiments as well as a reserve channel of communication via the Hq Ship.

The general outline airplan is:

A. <u>BOMBARDMENT ACTION</u>.

    1. <u>Preliminary bombardment</u>: Over the general area from D-21 to night of D-1, with concentration on WOOLACOMBE BEACH from D-7 to D-1. Bombing to cease at approximately H hour nightly. On the night D-1 WOOLACOMBE BEACH will be intensely bombed with the bomb best fitted for detonation of mines. The last wave of bombers will come in at H hour simultaneous with transports (see para D-1) and drop only blast bombs with delay fuses.

    2. <u>Saturation Bombing</u>: HE with a good percentage of white phosphorous smoke bombs to be laid on the area 91.5 - 66.0 to 94.5 - 64.5 to 93.0 - 61.0 to 91.5 - 64.0 to 91.5 - 66.0. This attack will commence not later than H plus 3 hour and cease at H plus 4 unless further attack is requested from the Hq ship. Destruction plus reduced visibility to impair artillery observation, is desired. Extreme care will be taken that <u>no</u> bombs fall East or North of the Southern R.R.

3. <u>Bombline</u>: From H minus 3 the bombline, except for specifically designated targets, will be road BERRYNARBOR - COMBE MARTIN - KENTISBURY - ARLINGTON and hence along RIVER YEO to BARNSTAPLE.

4. <u>Smoke</u>: <u>a.</u> Smoke bombs reinforced by SCI, will be laid from MORTE POINT to WOOLACOMBE and from WHITING HOLE to PICKWELL at H plus 3. Later reinforcement by either bombs or SCI, to be furnished as requested. The shoulders and shore line of CROYDE BAY to be similarly screened at the same hour.

    <u>b.</u> A smoke carpet will be placed over WOOLACOMBE BEACH and CROYDE BEACH at H plus 3 at a low altitude to protect the unloading of the LCTs and other craft from hostile air attacks. It will be reinforced upon request.

5. Bridges and Crossing over RIVER TAW and RIVER YEO: request bridges be previously destroyed, and crossings kept neutralized after H hour, from coast inland 10 miles.

B. FIGHTER ACTION

    1. <u>Fighter cover (and escort as required)</u> provided during daylight hours for:

        <u>a.</u> Passage of convoy
        <u>b.</u> Beachhead, particularly WOOLACOMBE BEACH and CROYDE BEACH
        <u>c.</u> Troop Carrier transport formations
        <u>d.</u> Bomber and cannon fighter formations as situation may demand

    2. <u>Bomber and cannon fighters</u>

        <u>a.</u> Alerted to attack such targets as reported in Para C 2 below
        <u>b.</u> Alerted to respond to approved requests from Hq Ship
        <u>c.</u> Sweeps, within the remaining capability of equipment, beyond the bomb line seeking for, and attacking, targets of opportunity.

C. RECONNAISSANCE

    1. Reconnaissance missions from H plus 3 hour on at 20 minute intervals until H plus 6 and thereafter on call from the Hq Ship. Preliminary reports will be rendered immediately to the H2 Ship by radio and full and complete reports upon landing at base airdrome.

    2. Tactical reconnaissance of roads and reserve movements towards this Division zone of action will be undertaken. Reports will be submitted to base airdrome immediately by radio, Hq Ship and Division Hq will both maintain a listening watch on the tactical reconnaissance frequency.

    3. Photographic missions as may be requested by the Div CG.

    4. <u>All</u> air units engaged will be alert to observe and report on landing enemy installations and activities.

D. AIRBORNE TRANSPORT

    1. Will be brought in on night D-1 with final bomber waves. Bombs dropped on final wave to be of blast type, thus permitting freedom of movement to the paratroops under their delayed action.

UNIT	DZ	TIME	DATE
101st A/B Div less Prcht Regt	Area BARNSTAPLE-BITTADON-ILFRACOMBE ROAD	H plus 3	D day
501st Prcht Regt and Prcht Regt., 101st A/B Div	- same -	H	D day

    2. Fighter escort, as provided as in Para B 1 (c) above, for transport formations during X daylight hours.

E. COORDINATED ATTACKS

    1. A coordinated airground attack will take place with Rangers on batteries at 89.0 - 68.0 and at 86.5 - 63.0 both at H plus 3. This attack

is liable to cancellation. Positive clearance prior to attacking targets must at all costs be obtained from Hq Ship. Orders will be given at this time to either:

    (a) Attack batteries

    (b) Attack alternate target

    (c) Attack specific target, or

    (d) Withdrawn withholding attack.

This attack will be staged in the following sequences:

A flight of strafing cannon fighters, to be instantly followed by a flight of dive bombers which will in turn be followed by a flight of dive bombers armed with a blast type of bombs of such characteristics and fuzesetting as to permit ground troops to advance under their delayed detonation. If previous Ranger attack has been successful a yellow and black panel will be displayed to indicate the battery is in our hands. This will constitute a safeguard for possible delay in communications from shore to Hq. Ship.

F. AA DEFENSE (presumably to be coordinated under an Air Defense Command).

    1. Friendly Type Aircraft: Types to be involved in the operation will be promptly reported to Ground Hq so immediate steps for concentrated instruction in identification recognition of those types may be instigated among the ground forces. Information of additional types added after the operation has begun will be transmitted to the Hq Ship.

    2. AA Guns: AA is given freedom of action to 2000 feet in convoy area. Friendly aircraft will accordingly not fly below 3000 feet except in pursuit of enemy aircraft in this area proper.

    3. Low Flying Aircraft: Low flying aircraft on target courses will be presumed hostile.

    4. Barrage Balloons: to be moored at Beach installations at such time as those installations and the balloons are secure from aimed fire.

    5. Mobile Aircraft Reporting Units: to be landed and installed as the Air Force Commander may desire.

    6. Mobile Operation Room Units (SCS-3): Same as para. 5 above.

G. LANDING GROUNDS

    1. Crash landing grounds: The most suitable landing lane and wind direction will be indicated by ground strips on SAUNTON SANDS.

    2. Emergency landing ground: The capture of CHIVENOR AIRDROME and establishment of operational landing lanes will be indicated by a triangle of ground strips as well as by radio message to Base Airdromes.

    3. Upon establishment of a proper airfield: At such time as a suitable airfield becomes operational, Servicing Rangers will undertake to furnish oil, petrol, and ammunition to fighter aircraft. Availability of this service will be duly reported to the Air Support Commander by the Force Commander.

H. GENERAL

    1. Major land demolitions: To be preceded 30 seconds by either a red or white flare to avert potential danger to low flying friendly aircraft.

2. **All support missions:** will check with the Hq Ship by radio before arrival over the target designated them and obtain final clearance before proceeding with their attack.

3. **Targets given aircraft in flight from Hq Ship:** such targets where aircrews have not had previous ground briefing will be clearly defined by easily visible landmarks. Targets submitted to AS parties will comply with this requirement.

SOLUTION TO FIRST REQUIREMENT PAR 4e.

Special equipment (not included in Table of Basic Allowances) required for this division and attached troops will be as set forth in Tables of Amphibious Equipment for Operations prepared by AFAF under date of 23 February 43 and revised to 1 June 43.

Solution to par #5 - requirement #1

1. Organization of landing beach.

   a. Typical plan is shown for Woolacombe beach - (similar layout can be adapted to Saunton Sands).

   b. Initial Plan: For beach exits shown on sketch #1. This work to be performed by two engineer companies (see Engr Plan, Sec III).

   c. Final Plan: for beach exits and beach layout shown on sketch #2. This work to be completed by the shore party engineers.

2. Organization of landing area back of beach

   a. Typical plan for Woolacombe area shown on sketch #3 (similar plan will apply to Braunton area).

   b. The organization of this area is based upon the following:

   (1) Initial reserves for assault regiments to be brought in with reserve battalion and assembled in Woolacombe and Braunton for distribution.

   (2) Initial assembly area for reserve battalion of assault regiments Woolacombe and Braunton.

   (3) Initial assembly areas in Woolacombe area for three combat teams of the follow-up division shown on the sketch 3. (A similar layout will apply to the Braunton area). Vehicles and personnel will move out from boats as soon as they land directly to these initial assembly areas where they will be organized and moved out at once to secure their initial objectives. Woolacombe will be used as an intermediate vehicle park for disabled vehicles. Disabled vehicles will be towed off the beach immediately to the intermediate park for repair and cleared to their proper CT areas as soon as possible.

   (4) Assuming that the follow up division pushes on to secure final bridgehead, ammunition, gas and oil, rations and water supply points will be established at Braunton and in area about 3000 yards due east of Woolacombe. To be established immediately after follow-up division clears the proposed supply areas. Sketch #3 shows these supply points for the Woolacombe area.

   (3) Engineer Plan

   Woolacombe area (similar plan for Braunton area)

   a. Clear and mark two 16 yard passages thru beach minefield from water line to beach exit at X (1 Company). Land immediately following assault infantry (probably H plus 30).

   b. Prepare road thru lanes cleared and maintain. (1 Company). Land at H plus 1:30. Will need about 500 yards landing mat bulldozer and picks and shovels.

   c. Remove obstacles, repair and maintain road from beach exits forward. (1 Company). Land at H plus 2:00. Bulldozers other road equipment.

   d. Prepare additional vehicle exit road 400 yards north of main exit, (1 Company to commence after completing task b.).

    e. Upon arrival of Shore party Engineer companies on a., b., and d. turn over tasks and assemble in front of Moelfre Hotel for further instructions.

    f. Shore Party Engineers to widen gaps in obstacles and clear an area 200 by 1000 yards by D plus 2 and 200 by 2000 by D x 4. Meantime troops landing must be moved at once to initial assembly areas well back from the beach, there they are checked and reorganized to move inland.

SKETCH No. 1
Solution to
Par. 5 Requirement I

BEACH TASKS FOR ENGINEERS WITH ASSAULT REGIMENT

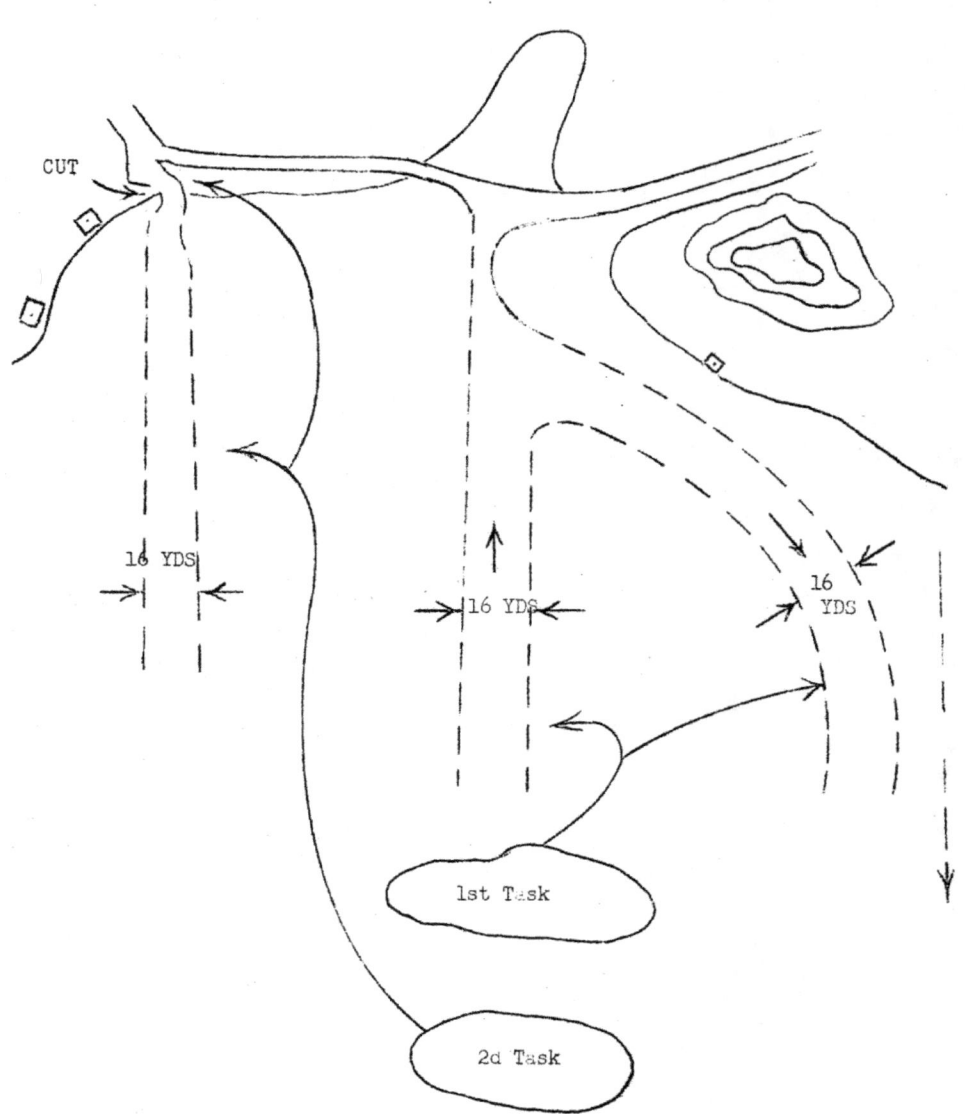

SKETCH NO. 2

Solution to Par 5, Requirement I

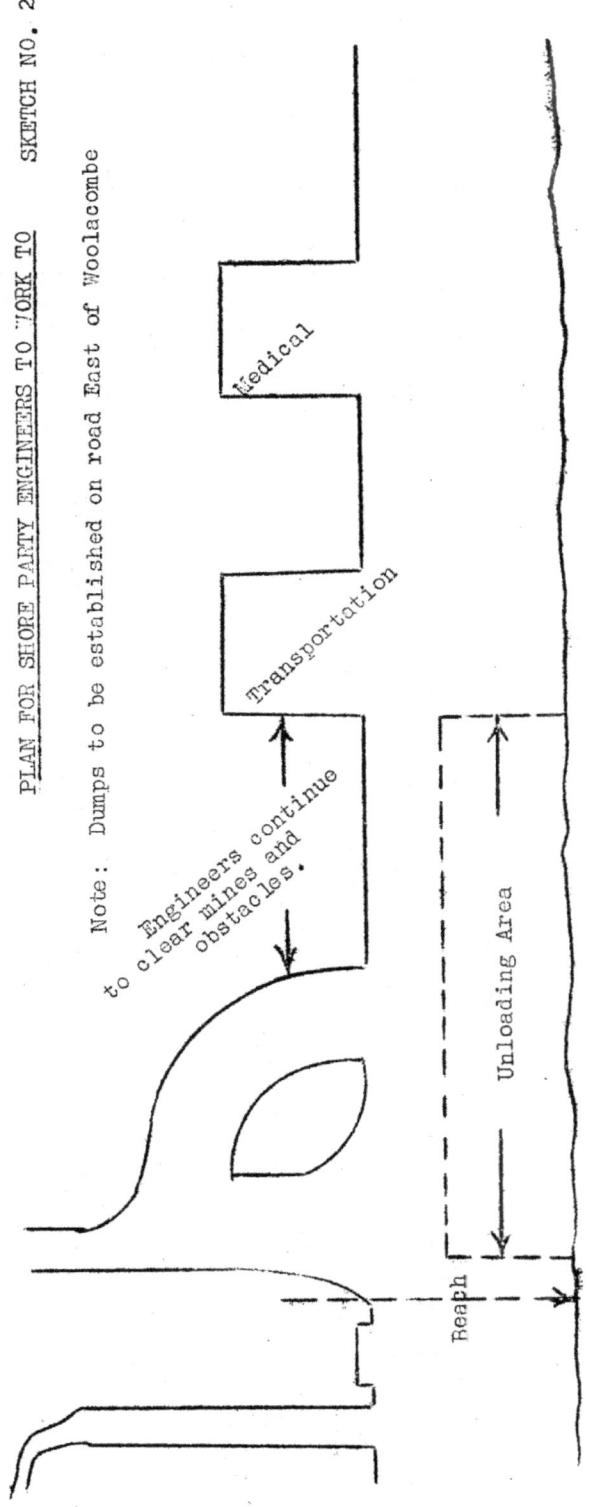

SOLUTION: FIRST REQUIREMENT, Para 5.

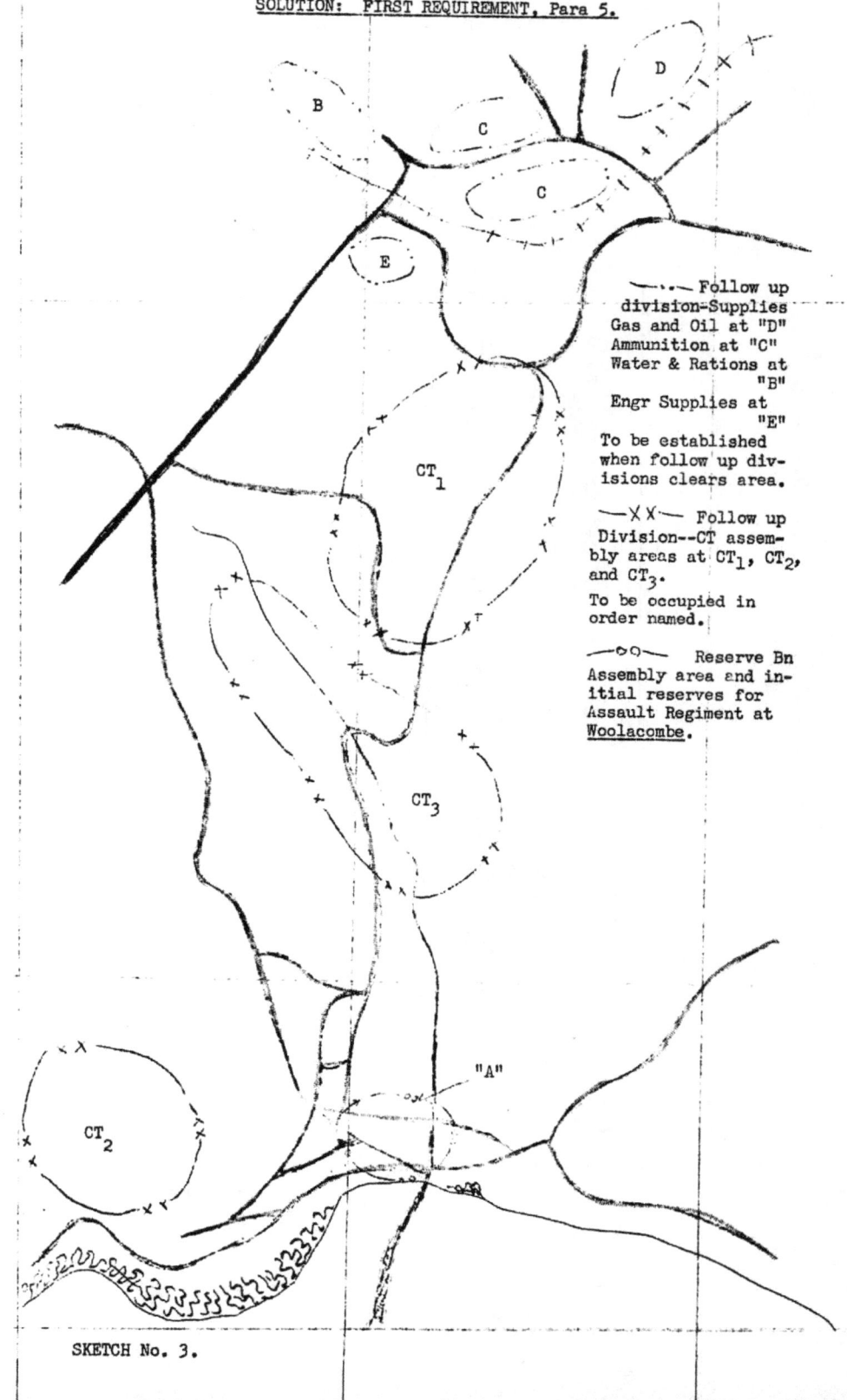

—··— Follow up division–Supplies Gas and Oil at "D" Ammunition at "C" Water & Rations at "B" Engr Supplies at "E"
To be established when follow up divisions clears area.

—XX— Follow up Division--CT assembly areas at CT_1, CT_2, and CT_3.
To be occupied in order named.

—OO— Reserve Bn Assembly area and initial reserves for Assault Regiment at <u>Woolacombe</u>.

SKETCH No. 3.

Solution to Second Requirement

2. CT 1 with the 3d Ranger Bn attached will land on Beaches in the Woolacombe area, immediately capture the high ground to the north and south of Woolacombe and then advance eastward to the Southern Railway and the highway running north thru INN ($\frac{1}{2}$ mi west of West Down) to ILFRACOMBE, where units will prepare for advance to the west bank of Colom Stream (see map).

Formation: 1st, 2d Bns and 2nd Rangers abreast from right to left
Landing Beaches: C.D.E. Boundary between BNCT (see map)

3. a. The 1st Bn CT will land at Beach "C" penetrate the hostile beach defenses near the boundary between Bns and the northern half of the enemy combat post defending the southern end of Woolacombe Beach and capture the high ground north of Rickwell Down. It will then seize the hill south of Willingcott and advance thereafter to the line of the Southern Railway - Dean. The 1st Bn will maintain contact with the 2nd Inf on the right.

b. The 2nd Bn Combat Team will land at Beach "D" and immediately capture Woolacombe and the high ground north and south thereof. It will then advance to the east and seize the general line of the DEAN - ILFRACOMBE road within the battalion zone of action and prepare without delay for further advance. The 2nd Bn will protect the left flank of the regiment.

c. The 3d Ranger Bn will land at Beach "E", establish contact with the 2nd Ranger Bn operating near Mortehoe. It will then advance southwest generally along the Mortehoe - Mortehoe Sta Road and assist the 2nd Bn in capturing the high ground near Mortehoe Sta, and assist the advance of the 2nd Bn CT to the line of the Dean - Ilfracombe Road and await further orders.

d. The 3d Bn Ct with the anti-tank company and the Cannon Platoon attached (Reg Reserve) will be prepared to land at Beach "C" or "D". One liaison group of 1 officer, 2 messengers and 1 radio operator equipped with radios will land with each assault Bn CT, establish close contact with them and facilitate employment of the Reserve CT in the zone of action of either assault Bn.

x. Bayonets will be fixed during hours of darkness. Weapons will be fired on orders of Platoon leaders only.

SOLUTION TO PAR 1 THIRD REQUIREMENT

Par 2 - the 2nd Bn 1st Infantry (reinforced) will land at Beach "D", penetrate the hostile beach defenses near the southern end of the golf links west of Woolacombe Down and immediately capture the high ground about ½ mile to the east. It will then capture hill 688 (northwest of Buttercombe), and seize the high ground west of the Southern Railroad and prepare for advance to the Division 2nd Objective.

    Formation: Cos E and F abreast, Co E on the right.

    Boundary: Between companies: Woolacombe (Town) - Woolacombe - Mortehoe - Mullacott Cross road (all to Co F)

    3. a. Co E will land on Beach "D" in its zone of action, immediately reduce weapons firing north along the beach from positions located near Potters Hill, capture Potters Hill, advance east in its zone of action and in conjunction with and assist the Ranger Bn and Parachute Troops assist Co F in capturing the high ground near Morthoe Station. It will then seize the ground west of the Dean - Mullacott Cross road and prepare for further advance to the east.

    Co F will land on Beach "D" in its zone of action, immediately reduce small arms weapons firing south along the beach from positions near Woolacombe and capture Woolacombe and the high ground to the north thereof. It will then advance to the east, capture the high ground near Morthoe Station, assisted by the 3d Ranger Bn and Airborne troops, advance to the northeast and seize the nose which runs north just west of the Mullacott Cross - Ilfracombe road. It will then protect the left flank of the battalion.

    Co G with one Platoon (Heavy MG) Co "H" attached will land on Beach "D" and be prepared to operate in the zone of advance of either Co "E" or Co "F". 1 officer or NCO and the runner of Co "G" will be assigned to each assault Co as a liaison detail and will be prepared to act as guides for Company "G" in the zone of action of the unit to which they are attached.

    Co H, less 1 Platoon (Hvy MG), will land at Beach "D" and be prepared to support the attack of either Assault Co. One liaison group of 1 officer and 1 messenger will be detailed as liaison to each Assault Co.

SOLUTION TO PAR 2, THIRD REQUIREMENT

LANDING PLAN AND DIAGRAM

Wave 1   8 LCVP's

◿ ◿

2d Platoon Co F	35		1st Plat Co F	35
60 mm Mort Sqd	5		60 mm Mort Sqd	5
L.M.G. Sqd	7		L.M.G. Sqd	5
Hq Mortars	5		Hq L.M.G.	2
Co Hq	2		Co Hq	7
Engrs	6 60		Engrs	6 60

◿ ◿

1st Plat Co F	35		2nd Plat Co E	35
60 mm Mort Sqd	5		60 mm Mort Sqd	5
L.M.G. Sqd	5		L.M.G. Sqd	5
Hq L.M.G.	2		Hq Mortars	2
Co Hq	7		Co Hq	7
Engrs	6 60		Engrs	6 60

◿

1st Plat Co E	35
60 mm Mort Sqd	5
L.M.G. Sqd	5
Hq L.M.G.	2
Co Hq	7
Engrs	6 60

Wave 2   4 LVCP's

◿

3d Plat Co F	35
60 mm Mort Sqd	5
Hq Wpns Plat	7
Co Hq	7
From Co G	2
From Co H	2
Medical Detch	2 60

◿

3d Plat Co F	35
60 mm Mort Sqd	5
Hq Wpns Plat	7
Co Hq	7
From Co G	2
From Co H	2
Medical Detch	2 60

◿

3d Plat Co E	35
60 mm Mort Sqd	5
Hq Wpns Plat	7
Co Hq	7
From Co G	2
From Co H	2
Medical Detch	2 60

Wave 3   4 LVCP's

◿

2d Plat Co G	35
60 mm Mort Sqd	5
L.M.G. Sqd	5
Hq Mort Sec	2
Engrs	8
Co Hq	5 60

◿

1st Plat Co G	35
60 mm Mort Sqd	5
L.M.G. Sqd	5
Hq L.M.G.	2
Co Hq	7
Engrs	6 60

Wave 4   4 LCVP's, 5 LCM(3)'s

◿

3d Plat Co G	35
60 mm Mort Sqd	5
Hq Wpns Plat	7
Co Hq	5
Medical Detch	2
Shore Party	6 60

▭ ▭ ▭ ▭

Light Tank Plat	29
Naval Detch	45
Shore Party Med	11
Shore Sig Det	18
Engr Co	100

◿

| Heavy Wpns Plat | 45 |
| Engrs | 16 61 |

-29-

Wave 5   6 LCVP's
         ◠ ◠ 81 mm Mortar Plat   62
              Equipment   28   90

         ◠ Hq Co H   30

Wave 6   4 LCVP's,
         6 LCM(3)'s
         ◠ ☐ Plat Cannon Co   34
             Platoon A.T. Co.   35
             Bn Hq Co   35
             Mine Squad   8
             Trans Sec Ser Co   12

         ◠ ◠ ◠ Hq & Hq Co 2d Bn   100
              Shore Fire Con Pty   8   12   120
              Equipment

         ◠ 2d M.G. Plat Co H   45
           Equipment   15   60

         ☐ ☐ ☐ Engr Co Shore Pty   93
               Arty Bty (Light)   114

Wave 7   5 LCV (5)
         from Far
         Shore
         ◠ ◠ ◠ ◠ ◠ Battalion Transportation & Equipment

-30-

SOLUTION TO FOURTH REQUIREMENT

2. Co "E" will land at Beach "D" with the 1st and 2nd Platoons abreast, the 1st Platoon on the right, immediately reduce weapons firing from positions near POTTERS HILL, and capture POTTERS HILL. It will then assist Co "F" in the capture of the high ground near MORTHOE STATION and advance EAST to the general line of the DEAN - MULLACOTT CROSS ROAD where it will immediately prepare for further advance. The Company will be prepared to assist the 1st Battalion by fire in the capture of HILL 688 - (1000 yards) NORTH of WILLINGCOTT and will coordinate that action with operations of the 101st Airborne Division from the EAST.

3. a The 2nd Platoon will land at Beach "D" on a 200 yard front and will advance to the EAST with its left resting generally opposite the SOUTHWEST EDGE of WOOLACOMBE. It will immediately locate and reduce hostile weapons firing NORTH from positions near the NORTH side of POTTERS HILL and assisted by the 1st Platoon capture POTTERS HILL. It will then advance to the EAST along the draw EAST of WOOLACOMBE, assist Co "E" in the capture of the high ground at MORTHOE STATION and await further orders.

b The 1st Platoon will land at the SOUTH of Beach "D" and with its right resting on the Battalion right boundary assist the 2nd Platoon in the capture of POTTERS HILL and then capture the high ground to the SOUTH as far as the SOUTH BOUNDARY of the Battalion. It will then advance to the NORTHEAST along the high ground parallel to the GEORGEHAM - WILLINGCOTT ROAD to a position near WILLINGCOTT where it will await further orders.

c The 3d Platoon will land in rear of the 2nd Platoon, clear the beach without delay, and follow the 2nd Platoon to the bend in WOOLACOMBE - MORTHOE STATION ROAD, 1000 yards NORTHWEST of WILLINGCOTT where it will await further orders.

SOLUTION TO FIFTH REQUIREMENT

2. The 2nd Platoon will land at Beach "D" opposite the south-west corner of Woolacombe at H-hour and will immediately locate and capture hostile weapons firing from positions near POTTERS HILL, capture POTTERS HILL, and then advance to the north-east with the left resting on the stream south of Woolacombe. It will assist CO E in capturing the high ground near Mortehoe Sta and await further orders.

3.   a. The 4th Squad upon landing will deploy on the right at once and will advance rapidly on the north-west side of Potters Hill, quickly locate hostile weapons and by evasive action close with and capture the Weapons. It will then advance to the crest of POTTERS HILL, capture the northern and await further orders.

    b. The 5th Squad upon landing will deploy on the left at once and advance on the knoll at the base of the northern end of POTTERS HILL. It will quickly locate the most favorable point in the hostile position and assault and capture the knoll. It will then assist the 4th Squad in the capture of POTTERS HILL by attacking from the north.

    c. The 6th Squad will land in near of and follow the 5th Squad. It will take up column formation and be prepared to protect the left flank of the platoon or going into action on the left of the platoon.

A SOLUTION TO SIXTH REQUIREMENT

    2. The 4th Squad will land at Beach "D" on the right of the Platoon, quickly locate and capture hostile weapons firing from the north western side of POTTERS HILL. It will then advance up the western side of POTTERS HILL to the crest, capture the northern end and await further orders.

    3. <u>On landing the Squad</u> will rapidly take up a Squad column formation preceded by the reconnaissance group.

    The reconnaissance group (consisting of four (4) men) will advance rapidly to the location of enemy obstacles opposite the northern side of POTTERS HILL where it will quickly locate the most favorable point through which the squad will pass and attack hostile weapons firing from positions near POTTERS HILL.

    The squad (less the reconnaissance party) will take position in rear of and conform to the advance of the reconnaissance party and will be deployed abreast of the reconnaissance group only when necessary to overcome hostile resistance.

ALL TIMES ARE B.S.T. LAT. 51° ½ N.) approximate.
METEOROLOGICAL DATA Long. 4° ¼ W.)

1943. DATE September.	NAUTICAL TWILIGHT BEGINS	CIVIL TWILIGHT BEGINS	SUNRISE	SUNSET	CIVIL TWILIGHT ENDS	NAUTICAL TWILIGHT ENDS	MOONRISE	BEARING	MOONSET	BEARING	AVERAGE NO. OF DAYS IN MONTH OF				
											CALMS	WIND 1 - 3	WIND 3 - 6	WIND GALE	RAIN	FOG
1st	0512	0553	0627	2004	2037	2118	0755		2058		1			1	14	3
2nd	0514	0555	0628	2002	2035	2116	0900		2121		Average % of N. Winds = 10%				
3rd	0516	0557	0630	2000	2033	2114	1006		2143		" " S. " = 10%				
11th	0530	0611	0643	1942	2015	2056	1829		0249		" " NE,E & SE " = 34%				
12th	0532	0613	0644	1940	2012	2053	1908		0406		" " NW,W & SW " = 44%				
13th	0534	0614	0646	1938	2010	2051	1940		0527		" " of calms = 2%				
14th	0536	0616	0647	1936	2008	2049	2007		0649		Average amount of cloud = 6/10				
15th	0538	0618	0649	1934	2006	2047	2024		0810							

PHASES OF THE MOON.

7th First Quarter   14th Full
21st Last Quarter   29th New

ALL TIMES ARE B.S.T.
METEOROLOGICAL DATA

B.S.T.

1943 DATE September.	NAUTICAL TWILIGHT BEGINS	CIVIL TWILIGHT BEGINS	SUNRISE	SUNSET	CIVIL TWILIGHT ENDS	NAUTICAL TWILIGHT ENDS	MOONRISE	BEARING	MOONSET	BEARING	AVERAGE NO. OF DAYS IN MONTH OF			
											CALMS	WIND 1-3	WIND 3-6	WIND 6-Gale
16th	0540	0620	0650	1932	2004	2045	2100		0928					
25th	0553	0633	0705	1910	1941	2021	0234		1749					
26th	0555	0635	0706	1908	1939	2019	0336		1816					
27th	0557	0637	0708	1906	1936	2016	0440		1841					
28th	0559	0639	0709	1904	1934	2014	0544		1903					
29th	0600	0640	0711	1902	1932	2012	0651		1925					
30th	0602	0642	0712	1900	1930	2010	0757		1948					

TIDAL DATA

ALL TIMES ARE B.S.T.

B.S.T.

1943. DATE September	HIGH WATER	HEIGHT FEET	LOW WATER	HEIGHT FEET	HT. OF TIDE AT TIME OF LANDING	DATE OF SPRINGS LAST	DATE OF SPRINGS NEXT	DATE OF NEAPS LAST	DATE OF NEAPS NEXT	BEACH GRADIENT AT H.W.	BEACH GRADIENT AT L.W.	TIME OF LANDING	REMARKS	REMARKS
1st	0746	27.4	0132	1.8		Springs								
	2004	28.1	1346	1.8										
2nd	0818	27.4	0202	1.4		Springs				Place			Average Charted Gradient between Chart Datum & H.W. Line	
	2032	27.8	1414	1.7						Woollacombe Sands			1/80) Probably steeper	
3rd	0848	27.1	0230	1.7				1		Croyde Bay			1/90) near H.W. Line.	
	2104	27.3	1442	2.0						Saunton Sands.			1/120)	
11th	0411	23.9	1023	4.9			-3							
	1637	25.4	2255	3.3										
12th	0509	26.0	1117	2.7			-2							
	1734	27.5	2344	1.3										
13th	0601	27.8	1206	0.8			-1							
	1823	29.1	-	-										
14th	0647	29.2	0032	0.2			Springs							
	1909	30.2	1253	0.6										
15th	0729	29.8	0118	1.1			Springs							
	1951	30.2	1336	1.3										

TIDAL DATA.

ALL TIMES ARE B.S.T.
METEOROLOGICAL DATA

1943 DATE September.	HIGH WATER	HEIGHT FEET	LOW WATER	HEIGHT FEET	HT. OF TIDE AT TIME OF LANDING	DATE OF SPRINGS		DATE OF NEAPS		BEACH GRADIENT AT		TIME OF LANDING	REMARKS	REMARKS
						LAST	NEXT	LAST	NEXT	H.W.	L.W.			
16th	0811	29.3	0202	1.3		1								
	2031	29.5	1416	0.8										
25th	0409	22.1	1021	6.8				3						
	1637	23.6	2249	5.3										
26th	0502	23.8	1106	5.0				4						
	1722	25.0	2329	3.6										
27th	0542	25.2	1143	3.5										
	1801	26.2	-	-			-3							
28th	0617	26.3	0004	2.4			-2							
	1834	27.3	1217	2.4										
29th	0648	27.4	0038	1.6			-1							
	1903	28.3	1249	1.6										
30th	0720	28.2	0110	1.2			Springs							
	1934	28.5	1318	1.0										

EXERCISE 2.

INDEX

EXERCISE "BREW"

	Page No.
Purpose	1
Introduction	1
General Situation	1
Estimate of the Enemy Situation	
Ground	1
Air	2
Mission of U.S. Task Force	2
Special Situation	2
Requirements	3
Order of Battle	
Ground	3
Air	4
Field Orders:	
First Division, FO #1	5
Second Infantry, FO #1	7
1st Bn, 2nd Inf, FO #1	9
Annex:	
I G-2	12
II Air	13
III Naval	
A Landing Ships and Craft	18
B Boat loading schedule	19
C Naval Gunfire Support	20
D Graphic schedule of Naval Gunfire Support	21
IV Fire Support Groups	22
V Solution to problem	
A General	23
B Organization	25
C Equipment List	26
D Far Shore Detachment	27

ASSAULT TRAINING CENTER
CONFERENCE
HQ ETOUSA

11 June 1943
KM

EXERCISE "BREW"

PURPOSE:

1. The purpose of exercise BREW is to study the problems which confront an Assaulting Force on a heavily defended shore.

INTRODUCTION:

2. Activation orders have been received from Allied Force Headquarters constituting a U.S. Task Force, which is to engage in an invasion of a hostile shore, from bases situated in England. The invasion is to be referred to by its code designation, CHASEM.

3. <u>The mission of the U.S. Task Force</u> - This Force is part of a coordinated invasion of Devon. FORCE GUARDS (British) carrying out a landing operation north and inclusive of 35 miles simultaneously with operations of U.S. Task Force. At the same time landings will be made by other troops on the south coast of the mainland.

4. Immediately upon receipt of the above orders the Commanding General, U.S. Task Force, summoned his complete staff and outlined to them the general plan, at the same time charging them with the responsibility of preparing the detailed plans necessary to accomplish his mission.

5. Invasion is to be launched from England where U.S. Task Force has been engaged in extensive precombat training. The ports and beaches made available have been kept to a minimum to further demonstrate the problem of initial supply, and the problem of supplying an invading Force as the assault progresses.

6. D-day is 15 June 1943. H-hour is 0210 hours.

GENERAL SITUATION:

7. The west coast of DEVON is assumed to be a part of GERMAN OCCUPIED FRANCE. The coast is further assumed to be as shown on the map between MORTE POINT and HARTLAND POINT (both inclusive), except that the coast is assumed to run due north from the eastern edge of MORTE POINT (ROCKHAM BAY) and due south from HARTLAND POINT.

8. Except for the geographical location of the DEVON coast and the direction of attack of our forces, this coast may be considered to be actually part of the French coast.

9. The attack on the coast will be a combined BRITISH and AMERICAN assault, each force having a corps of four (4) infantry divisions and one (1) armored division, all reinforced. The boundary between the BRITISH and AMERICAN forces is assumed to be a line running due east and west from ROCKHAM BAY (MORTE POINT). The AMERICAN corps, on the south, will be referred to as "Task Force A".

ESTIMATE OF THE ENEMY SITUATION:

10. The entire coastline is heavily fortified. All feasible sea approaches are obstructed and mined in accordance with the German doctrine. Beaches are obstructed and mined (see Intelligence Overlay). Gun emplacements are adequately and

continually manned. In certain instances, dummy guns have been installed. Intense light flak may be expected along entire coastline, intense heavy flak at strategic points, intense light and heavy flak and searchlights may be expected at all ports and major cities. Intense light and heavy flak may be expected at all airdromes, major supply dumps, railheads and important truckheads.

11. The enemy holds the coast (in both directions from the BARNSTAPLE BAY area) in considerable strength, divisions holding sectors measuring about 25 to 35 miles in width. All sectors are fortified on about the same scale as the BARNSTAPLE BAY sector.

12. A panzer division is located in the EXETER area. This division can close in the BARNSTAPLE BAY area within 24 hours after the beginning of an assault. Another panzer division can be expected to close in the area within 48 hours of an assault.

13. The enemy's defensive plan is based on holding the line of the coast. It consists of the following elements:
    <u>a</u>. Artillery fire and air attack to prevent the approach of craft to landing points.
    <u>b</u>. Fire of artillery, infantry weapons, antitank guns, etc, to destroy troops on the beach, while their egress from the beach is being prevented or slowed down by obstacles.
    <u>c</u>. Mobile reserves to destroy any troops penetrating the defenses of the immediate shoreline.

ENEMY SITUATION - AIR:
14. The enemy has at his disposal in this area 600 first line aircraft as follows:

    Fighters --- --- --- 300
    Night Bombers --- --- 100
    Medium & Light --- --- 50
    Fighter Bombers -- --- 150

disposed among aerodromes as indicated on the ordnance survey of GREAT BRITAIN - Air - ($\frac{1}{4}$" = 1 mi).

MISSION OF U.S. TASK FORCE:
15. The mission of the U.S. Task Force will be to establish a beach head sufficiently deep to remove all medium artillery fire from the beaches of such depth (10,000 -12,000 yards) as to enable the build up divisions to land full scale and assemble preparatory to carrying out their missions.

16. The assault will be made with two divisions abreast, which will be followed in by two immediate follow up divisions, with the armored division for exploitation.

17. The boundary between assault divisions will be the TAW RIVER, the first division on the north and second division on the south.

18. The objective of the Task Force will be to secure a line approximately along the ILFRACOMBE - BARNSTAPLE road continuing generally south from BARNSTAPLE.

SPECIAL SITUATION:
19. It is assumed that the Task Force order has designated the 1st Division as the northern assault division of the corps. This division is to be landed with assault scale equipment and vehicles, and the units are to be stripped of all administrative personnel with only sufficient drivers for actual vehicles carried. The balance of the 1st Division will land after the

assault follow-up Division. This division will have a minimum Shore Party, not to exceed 1,500 men. The assault follow-up Division will pass through this same area and land light scale. The balance of the Shore Party with attachments will land with the follow up Division and augment the advance Shore Party Group with 1st Division and organize the beaches in the area.

20. An Airborne Division, 2 Ranger Bns and a Parachute Bn will support the operation of the 1st Division in the assault. Naval gunfire support will be provided by the covering Naval Force Division and the Division will be aided by special support craft.

21. In addition to the Shore Party mentioned above, the Division will have the following attachments: 1 Bn Combat Engrs, 1 Tank Destroyer Bn, 2 AA Bn (AW) (SM), 1 Tank Bn, and 1 CWS (Smoke) Bn.

REQUIREMENT

22. Prepare and coordinate Divisional, Regimental, Battalion and Air Force Orders (for one of the Battalions, and Air Support).

23. Give the composition arms and equipment of an Assault Battalion and of all units down to and including squads.

ORDER OF BATTLE - GROUND:
24. Inf Rgt (less Adm Personnel and drivers except for vehicles assigned)
    1 Bn Combat Engineers
    1 Bn FA (3 batteries,. Each battery will have 6 SP
        Guns mounted - 3 guns per LCT(5))
    1 Medical Collecting Co.
    1 Far Shore Bn
    1 AA Bn
    1 Smoke Co, CWS
    1 Tank Co

25. Assault Bns
    1 Rifle Co (Light assault armed and equipped)
    2 Rifle Co normal (less adm)
    1 HW Co
    1 Co Combat Engineers
    1 Plat Tanks
    1 Plat Smoke Co CWS
    Detachment of Shore Party
    Shore Fire Control Party
    Air Support Party (One per Div Hq, and one per initial
        assaulting battalion)

26. Res Bn
    Normal Bn
    1 FA Battery
    1 Med Collecting Co
    1 Platoon Tanks

27. Regt Group
    Bal of Regt & attachments.

ORDER OF BATTLE - AIR:
28. 9 Medium Bmbr Groups consisting of:
         4 Groups B-25 (25 a/c per sqd)
         5 Groups B-26 (19 operational)
     4 Light Bomber Groups A-20 (19 operational)
     6 Dive Bomber Groups A-36 (25 a/c per sqd)
                              (19 operational)
     13 Fighter Groups
         7 Groups P-47 (25 a/c per sqd)
         6 Groups P-51 (19 operational)
     1 Night Fighter Group
     9 Troop Carrier C-53 (13 C-53 & C-47 per sqd)
              C-47 (10 operational)
     2 Reconnaissance and Observation Groups P-51
     1 Photographic Group

ASSAULT TRAINING CENTER
CONFERENCE
HQ ETOUSA

HEADQUARTERS FIRST DIVISION

Aboard Headquarters Ship
"VULTURE"
1300 Hours 14 June 1943

Field Order #1

Maps: Devonshire 1/10560

1. a. Enemy situation (See G-2 Annex and Overlay).
   b. This Division is a part of Task Force "A" which is making the U.S. effort in an assault on the coast of FRANCE and is part of a combined BRITISH and UNITED STATES Force. The BRITISH sectors extend northward from a line running from the north side of MORTE POINT. The Task Force "A" is a corps consisting of four reinforced infantry division and one armored division. The assault of Task Force "A" will be made on a 2-division front. The U.S. northern boundary is the southern boundary of the BRITISH Corps. Its mission will be to establish a beach head of sufficient depth to remove observed artillery fire from the shore so as to enable the build up (reserve Divisions) to land full scale.

   Naval fire support (see Naval Annex)

   Air Support (see Air Annex)

   Parachute Support (see Air Annex)

   Airborne Support (see Air Annex)

2. a. This Division will land on Yellow, Blue and Red Beaches with the objective of securing a line generally along the high ground to the west of the L & SW Railway within the division area north of the TAW RIVER, and south of the BRITISH sector (see Opns Overlay). The assault battalions will over-run the beach defenses and take the initial objective (see Overlay). After the Reserve Battalions of the Assault Regt land they will form along the line of the intermediate objective preparatory to launching an attack against the division objective.
   b. Direction of attack (see Opns Overlay).
   c. Time of landing: H hour 0210 June 15 - (First Refm 0150 GMT) - (See Landing Schedule).
   d. Zone of action (See Opns Overlay).
   e. Reinforcements and attachments
      (1) Shore Engineer Group, consisting of:
          Group Hq and Hq Co, Naval Component,
          3 Bns Engineers, 1 Medical Det (Amphibious),
          1 Beach Signal Co, 1 MP Det, 1 Chemical
          Decontamination Co, - Total - 1,500 men
      (2) 1 Bn Combat (Beach Assault) Engineers -
          Total 600 men.
      (3) 1 Tank Destroyer Bn - Total 500 men
      (4) 2 AA Bn (AW) (SM) - Total 1,000 men
      (5) 1 Bn Tank (Med) - Total 500 men
      (6) 1 Amphibious CWS (Smoke) (3 Co) Bn - Total 500 men

          Total: 3,100 men

In addition there will be 2 Ranger Bns in support.

3. a. The First Infantry will land one Bn on Yellow Beach and one Bn on Blue Beach with a Bn in mobile reserve. (See Opns Overlay). The Second Infantry will land in a column of battalions at Red Beach (see Overlay). The Tank Destroyer

Bn will be attached to the 2nd Infantry and will support the 2nd Infantry on their attack on BRAUNTON and thereafter to the south.

The zone of action (see Opns Overlay).
Time of landing (Refer to Landing Schedule).

b. Field Artillery Bns of 3 Batteries each with 6 SP Guns per Battery (embarked in 2 LCT(5) each will be prepared to fire from landing craft in support of the assault regiments. (See Landing Schedule and Composition of RCTs).

c. Aviation (see Air Annex)

d. Antiaircraft Artillery (see Landing Schedule)

e. The Tank Destroyer Bn will land at Red Beach (see Landing Schedule and Opns Overlay) and support the 2nd Infantry in their attack on their final objective.

f. Tanks (see Landing Schedule and Composition of RCT and BLT).

g. Engineers (see Landing Schedule and Composition of RCT and BLT).

h. The 3rd Infantry will be in reserve (see Landing Schedule).

i. (1) Ranger Bns - The 1st Ranger Bn will land on MORTE POINT at H-3 hour with the initial objective of capturing the coast defense battery on this point. Its intermediate objective will be to hold MORTE POINT during the main assault landings. It will then proceed to the final objective (see Opns Overlay). Contact will be maintained with the BRITISH Forces on the north.

(2) The 2nd Ranger Bn will land on BAGGY POINT at H-3 hours with the initial objective of capturing the Coast Defense Battery. Its intermediate objective will be to hold BAGGY POINT until the landing of the main body when it will assist the 1st Infantry in landing. It will then be attached to the 1st Infantry.

k. Parachute Bn - (1) - A Parachute Bn will be landed at the objective of its Task Force about 5.3/4 miles east of the top of MORTE POINT with the objective of attacking the heavy gun emplacements along the eastern most part of the Division objective. (See Annex II, Air Plan).

l. Airborne Division - This will be a Task Force unit and will land along the ILFRACOMBE - BARNSTAPLE Highway. This Airborne Division will have the mission of forming a defense position along the Corps objective to immobilize the enemy reserves to the east.

x. (1) Composition of Combat Teams (see Annex "A")
(2) Organization of Beaches (see Far Shore Group Annex).

4. (See Administrative Annex).

5. Radio silence except on Div order SOI No.3 in effect H-12 (SOI No.3 not prepared but assumed for purposes of problem.)

ASSAULT TRAINING CENTER
CONFERENCE
HQ ETOUSA

HEADQUARTERS 2ND INFANTRY
1st DIVISION

Aboard Headquarters Ship
"DOLPHIN"

Field Order #1

Maps:

1. a. The area along the DEVON coast is occupied in considerable strength by GERMAN forces. The beach area immediately above high water has been wired, mined and otherwise fortified for defense. Reconnaissance shows that there are some underwater obstacles but it is not considered that this will prevent landing. We have air superiority over the place of landing.

b. The first infantry division reinforced has the mission of establishing a beach head in its area of sufficient depth to remove observed artillery fire from the shore and to enable follow up divisions to land. The 2nd Infantry RCT will land at Red Beach on D-day and seize the high ground to the east of the beach, secure the initial assault line reorganize an advance to the intermediate objective where an attack will be launched against the division final objective. The first infantry RCT will land to the left (north) of the second infantry and the southern boundary is the division boundary (TAW RIVER).

c. Naval Fire Support - This regiment will be supported by naval fire support from various support craft (see Landing Schedule).

d. This regiment will be reinforced by 1 Bn Field Artillery, 1 Medical Collecting Co, 1 AA Bn, 1 Far Shore Bn, 1 Tank Co.

2. a. This regiment will land on Red Beach with the final objective of securing a line generally along the L & SW Railway within the regimental area, north of the TAW RIVER (see Operations Overlay). The landing will be made in a column of battalions, the first battalion in the initial assault, the first and second battalions will be landed in small craft and the third battalion will be landed in larger craft. After the initial objective has been reached the first battalion will continue the attack within its sector, and the second battalion will attack to the right of the first battalion within its sector. The 3rd battalion will be in reserve (see Operations Overlay).

b. Embarkation Table (see Annex #1).
Operations Overlay (see Annex #2).
Boat Assignment Table (see Annex #3).
Landing Schedule (see Annex #4).
H-hour 0210, 15 June 1943.
Composition of Forces (see Annex #5).

3. a. The first battalion will land on Red Beach with two companies in the assault and one in support. After overcoming the beach defenses it will advance rapidly to the initial objective and hold this line until the second battalion has advanced within supporting distance. It will then reorganize along the line of the initial objective and advance rapidly within its sector to the intermediate objective. The second battalion will

follow the first battalion from Red Beach. When it has reached the line of the initial objective it will reorganize and push rapidly forward within its sector to the east and north with the mission of enveloping and attacking the beach defenses along BRAUNTON BURROWS.

    <u>b</u>. The third battalion will be in the reserve initially and after landing will occupy the high ground within the area of the initial objective.

    Field Artillery - Field Artillery will initially be under control of the Regimental Commander and will land upon call (see Landing Schedule).

    Tank Destroyer Bn - Tank Destroyer Bn will land on Red Beach after the second battalion has cleared the beach. One company will then be attached to the second battalion and the battalion (less one Company) will be in reserve under regimental control.

    Tank Battalion - One Company will be attached to each battalion. The Tank Company attached to the first battalion will be in small craft and the balance of the Tank Battalion will be carried in large craft (see Landing Schedule).

    CWS Co - One platoon will be attached to each of the Rifle Companies of the first battalion.

    Shore Party Battalion - (see Boat Assignment Table and Landing Schedule).

    4. (See Administrative Annex).

    5. (See Signal Annex).

ASSAULT TRAINING CENTER
CONFERENCE
HQ ETOUSA

1st BATTALION 2nd INFANTRY
1st DIVISION

Field Order #1

1. a. The area along the DEVON coast is occupied in considerable strength by GERMAN forces. The beach area immediately above high water has been wired, mined and otherwise fortified for defense. Reconnaissance shows that there are some underwater obstacles but it is not considered that this will prevent landing. We have air superiority over the place of landing.

b. First infantry division reinforced will attack the coast of DEVON and seize and hold the high ground along the division objective. The second infantry will land on Red Beach in a column of battalions, and after over-running the shore defenses they will advance to the final objective and seize and hold this area.

2. a. The first battalion with the following attachments:
1 Co Combat Engineers
1 Platoon Smoke Co
Det of Shore Party and Shore Fire Control Party

will embark in small landing craft from APA #17 during darkness on D-day and will effect a landing twenty minutes after first light on Red Beach. It will clear the beach of small resistance and cut lanes from the beach defenses, and will then drive rapidly to the line of the initial objective, whereupon it will reorganize and continue the attack to secure the final objective within its zone of action protecting the left (north) flank of the second infantry and maintaining contact with the first infantry to the north.

b. Embarkation Table (Annex #1).

c. Operation Overlay (Annex #2).

d. Boat Assignment Table (Annex #3).

e. Landing Schedule (Annex #4).

f. H-hour - D-day 0210, 15 June 1943.

g. Composition of Forces.

3. a. Co A

Attachments: (see Composition of Forces) will land at H-hour D-day on the left (north) end of Red Beach on a front of 500 yards with two platoons in assault. It will rapidly clear the beach cutting lanes through beach defenses and envelop the enemy defenses along the high ground at the left of the sector. It will then push rapidly forward towards the east within its area (see Operations Overlay), and will support the advance of B Co on its left. After reaching the initial objective it will reorganize and be prepared to advance rapidly to the intermediate objective.

__b__. Co B

Attachments: (see Composition of Forces). Co B will land at H-hour D-day on Red Beach to the right (south) of Co A beginning at a point 500 yards south of the north end of Red Beach with two platoons in assault. It will rapidly clear the beach, cut lanes through the enemy defenses and attack to the eastward from all available covered positions until the line of the initial objective is reached. It will then reorganize and be prepared to advance eastward within the first battalion area.

__c__. Reserve: Co C will be in support and will land on Red Beach in the zone of both Co's A and B and will be prepared to support the attack of either company. After the beach defenses have been over-run this Company will make a flanking movement towards the south and envelop and capture the enemy positions along the BRAUNTON BURROWS from the rear in order to remove the threat of the enemy position in this area within the line of the initial objective.

__d__. Co D - Co D will not operate as a unit, personnel having been attached to various squads of the Rifle Companies (see Boat Assignment Table).

__e__. Co A 1st Combat Engineer Bn - One platoon of this company will be attached to A & B Co's to aid in the assault of the beach defenses.

__f__. Chemical Warfare Service Smoke Detachment (3 sections). One section will be attached to each of the Rifle Companies.

__g__. Shore Party Detachment (See Boat Assignment Table).

__h__. Shore Fire Control Party (see Boat Assignment Table).

__x__. (1) The SOP for debarkation, leading boat waves, etc, of the Second Infantry First Division will be strictly adhered to.

(2) Prompt landing reports will be made.

(3) The use of toxic chemicals by the enemy will be reported immediately.

(4) Liaison will be maintained with the unit on the left.

(5) Gas Sentries will be posted in each boat.

(6) Life belts will be discarded near the beach if possible in order to designate the route of advance of the leading elements.

4. __a__. Class I
   (i) Rations - All personnel will carry rations as issued. No further rations will be issued for the first 36 hours.

   (ii) Water - One canteen of water will be carried by each man.

__b__. No water from local sources will be used for drinking purposes.

__c__. No Class II supplies will be issued.

    <u>d</u>. Class III - None

    <u>e</u>. Class IV - None

    <u>f</u>. Class V - None

    <u>g</u>. Battalion Aid Station will be located in SAUNTON.

    <u>h</u>. Prisoners will be used to clear the enemy minefields.

  5. <u>a</u>. (1) When the first wave of each company has touched down on the beach the following signals will be used:

        Voice Code: NEWYORK
        Radio Signal: QKL
        Pyrotechnic Signal: Green Star Cluster

    (2) When the beach defenses have been over-run and the high ground 200 yards from the edge of the beach has been obtained the following signals will be given:

        Voice Code: FORT HAMILTON
        Radio Signal: XRB
        Pyrotechnic Signal: Red Star Parachute

    <u>b</u>. CP's
    (1) 1st Battalion - SAUNTON
    Co A initially in the quarry north of the hotel on SAUNTON SANDS. Advance with column in company zone of action.
    Co B initially in BOUNDARY DRAIN BM 18.1 and advance with column in zone of action.
    Co C with first battalion.

    (2) CP's will be reported promptly.

    <u>c</u>. Signal Communications - Access to signal communications will be SAUNTON SANDS - SAUNTON - Quarry north of FAIRLINCH - quarry norward of BRAUNTON.

ASSAULT TRAINING CENTER
CONFERENCE
HQ ETOUSA

HEADQUARTERS FIRST DIVISION

9 June 1943

Annex I to Field Order #1

G-2 ANNEX

1. a. The enemy holds the entire coast in both directions from the Barnstaple Bay area in considerable strength. Divisions holding sectors covering about 25 - 35 miles in width.

b. Defensive organization (see overlay).

c. Panzer division is located in the Exeter area. This division can close in the Barnstaple Bay area within 24 hours after the beginning of an assault. Another Panzer Division can be expected to close in the area within 48 hours of an assault.

2. a. The enemy is on the alert along the entire coast because of the strategic bombing that has taken place during the past two weeks.

b. (1) Antiaircraft artillery has been active.
(2) There was moderate fighter opposition until the last 48 hours. The Air Force now reports that it has obtained superiority over this section of the coast.

c. Other units are disposed as shown in the overlay.

d. No enemy movements or changes have taken place within the last 48 hours.

3. a. Enemy morale has suffered to some extent from the heavy strategic bombings during the last two weeks.

b. Weather is favorable for an assault.

4. a. The enemy is basing his plan of defense upon the fortified shore line with a mobile reserve situated far inland to reinforce any threatened break through.

ASSAULT TRAINING CENTER
CONFERENCE
HQ ETOUSA

Annex II to Field Order #1

Air Plan

GENERAL PLAN:
1. The 20th Tactical Air Force will support the operations of U.S. Task Force using the combat air units shown in the order of battle. This support will be coordinated with three landing phases, i.e.

      Phase I   (Assault Phase) H plus 6 hours
      Phase II (H plus 6 to D plus 2 inclusive)
      Phase III (D plus 3 to D plus 6 inclusive)

2. The Home Bomber Command will carry out bombardment missions on enemy lines of communications and strategic targets in Devon prior to and during the operation.

3. The Home Fighter Command will carry out fighter sweeps and escort bombardment missions. It will also furnish fighter cover, including night fighters, for the operation from "H" hour until the morning of D+2 days.

COMMUNICATION AND COMBAT:
4. Air Support Parties - (1) Air Support Parties to each Division Headquarters and each initial assaulting battalion of the Division.

(2) The Air Support Parties will embark with the units to which they have been assigned.

(3) Upon landing of the Division Air Support Parties the Battalion Air Support Parties will be recalled as seen fit by the Division Commander.

(4) The control of aircraft will be exercised by the Commanding General, 20th Tactical Air Force from Headquarters Ship from H hour until Command Headquarters and an Air Support Control is set up ashore.

5. Fighter Control - (1) Aircraft including night fighters, of Fighter Command supporting the operation will be controlled by Fighter Commander from Home based or Fighter Control Ship.

6. Air Warning System - (1) Complete Air Warning System composed of units of Air Warning Wing attached to 20th Tactical Air Force will be set up in Devon.

(2) Units for initial Air Warning System will be brought in over each sector on or about H+4 and immediately put into operation.

(3) Remainder of Air Warning System units will be brought in during the follow-up phases so that complete service will be provided.

(4) Prior to operation of assaulting Air Warning System control the assaulting reporting agencies will report into Fighter Control Ship and Home based Fighter Command.

7. __Paratroops__ and __Airborne__ - (1) One company of Paratroops will be used to attack and capture a heavily defended gun position near the force objective.

   (2) One Airborne Division will be dropped along Force objective to establish a line of resistance against movement of enemy reserves. These units will be reinforced through the beaches.

8. __Smoke__-- Three squadrons of smoke laying aircraft will be allotted the Division to facilitate the smoke plan. One squadron will be held in reserve on call for unforeseen mission.

9. __Narrative Plan__:
   Operation   prior to D day.
   (1) 20th Tactical Air Force
       (a) Carry out planned observation and reconnaissance missions.
   (2) Home Bomber Command
       (a) Carry out planned bombardment missions
   (3) Home Fighter Command
       (a) Carry out fighter sweeps
       (b) Provide fighter escort for missions of VIII Bomber Command.

10. __Operation during Phase I__ (H through H plus 6 hours)
    (1) 20th Tactical Air Force
        (a) Air Support Parties will land with assaulting troops of battalions at H plus 30 and prepare to transmit request for air support.
        (b) The Airdrome Squadron land at H plus 4 hours to conduct initial operation on captured airdrome.
        (c) Air Warning System Units land at H plus 4 hours and establish Air Warning System.
        (d) Engineer Battalion (Aviation Battalion) land at H plus 4 hours and move to captured airdrome, to prepare field for operation.

    (2) VIII Fighter Command
        (a) Furnish fighter cover for convoy and landing operations.

11. __Operations during Phase II__.
    (1) Tactical Air Force
        (a) Carry out planned bombardment missions on completion of which the aircraft will be available for request missions.
        (b) Carry out planned observations and reconnaissance.
        (c) Deal with smoke plan holding smoke aircraft in reserve for call missions until released by Air Force Commanding Officer.
        (d) Coordinate airborne plan.
        (e) Two Squadrons of Dive Bombers assigning one squadron each initial landing area. One squadron being held in reserve. One flight per Air Support Party will be maintained on air alert and controlled directly from Air Support Party to aircraft. After H plus 10 hours control of these aircraft will revert to Air Support Control.
        (f) Fighter aircraft of the Tactical Air Force and remainder of bombardment aircraft will be available for request missions through Air Support Control.
        (g) Observation planes will be available for missions on request of Force Commander or Commander Tactical Air Force.

(2) Home Fighter Command (until morning D plus 2 days)
    (a) Protect convoy and beaches
    (b) Protect assault shipping
    (c) Protect ports when occupied
    (d) Maintain air superiority over zone of operations of U.S. Task Force.
    (e) Furnish and control Night Fighters.

ASSAULT TRAINING CENTER
CONFERENCE
HQ ETOUSA

Annex II - AIR

DETAIL AIR PLAN DURING INITIAL ASSAULT

D-1
23.10   1. High level Strat. Bomb to be stopped on assault area at H-3.

D
02.00   2. 1 sqd A-20 to attack targets 6, 7, & 8, with
02.05   gunfire and para-frag bombs low level alt.

02.00   3. Fighter cover patrol beach area. 3 sqd 1-P-47, 2 P-51 to be increased or decreased as enemy situation allows.

02.05   4. 1 Flight B-26 attack target 4, preceded by gun attack from 2-P-51.

01.55   5. Smoke in plan A to be placed so to cover from MORTE POINT south to RIVER TAW, smoke screen.

02.10   6. Smoke bombs on MORTE POINT, BAGGY POINT, and SAUNTON
3.25   DOWNS, to be maintained.

02.10   7. 1 Sqd B-25 target 5 and area.

02.30   8. 1 Sqd B-26 target 3.

02.25   9. 4 sqd P-51 used in close cover support of mainland advance.

02.45   10. Tactical Reconnaissance of roads Bridgehead area.
onward

02.30   11. 4 Med, 2 light, at 60 min. notice.
onward

02.35   12. 2 Dive bomb sqd assist Rangers on target 1 and 2 if necessary.

01.30   13. Dropping of one bn Paratroopers to attack target area f on dropping zone 972660, 2½ miles from target.

02.00   14. Dropping of one Airborne Div along line of road from 972660 to road junction 980610.

ASSAULT TRAINING CENTER
CONFERENCE
HQ ETOUSA

Annex II - AIR

BOMB & MACHINE GUN TARGETS IN BEACH AREA

Subject to air attack during assault

1. 4 fixed coastal guns
   1 light AA
   892   675

2. 4 fixed coastal guns
   1 light AA
   865   626

3. Headquarters
   705   630

4. 7 Heavy AA
   700   644

5. 2 Hv Railroad guns
   928   639

6. 3 Light machine guns
   1 AT
   884   629

7. 2 AT
   1 LMG
   893   599

ASSAULT TRAINING CENTER
CONFERENCE
HQ ETOUSA

Annex IIIA to Field Order #1 (Naval)

LANDING SHIPS AND CRAFT

Type	Assault Regt (2)		Reserve Regiment and Div Troops	TOTAL (Divisions)	
	Assault Bn (4)	Res. Bn (2)			
APA	1			4	APA
(LCVP)	(30)			(120)	LCVP
(LCM(3))	(3)			(12)	LCM(3)
(LCS(S))	(2)			(8)	LCS(S)
LST		4	12	20	LST
LCI(L)		4	12	20	LCI(L)
LCT(5)	6	4	12	44	LCT(5)
LCT(R)	1			4	LCT(R)
LCG	4			16	LCG
LCF	1			4	LCF
LCC	1			4	LCC
LCS(L)	2			8	LCS(L)

ASSAULT TRAINING CENTER
CONFERENCE
HQ ETOUSA

Annex IIIB to Field Order #1

BOAT LOADING SCHEDULE

I. ASSAULT WAVE, WAVE "A"

    A. LCVP:

        1. Personnel: Officers - 1, EM - 29, including: 1-Officer, 16-Riflemen, 8-Assaultmen, 2-Crewmen LMG, 3-Crewmen 60 mm Mortar.

        2. Armament : Rifle, Cal 30, 03-1; Rifle cal 30 M-1 - 15; Carbine, Cal 30 - 13; BAR, Cal 30 - 1; LMG, Cal 30 - 1; Mortar, 60 mm - 1.

    B. LCM

        1. Personnel: Em - 11, including: 5 - Tank Crew (for Mark IV Tank), 6 - EM (Engr Dem Party).

    C. TOTAL PERSONNEL: (12 LCVP, 6 LCM)

Officers:	12 x 1	12
Riflemen:	12 x 16	192
Crewmen LMG;	12 x 2	24
Crewmen Mor:	12 x 3	36
Assaultmen:	12 x 8	96
Infantry - Total		360
Engr Dem Party:		36
		396

II. SUPPORT WAVE

    A. LCVP Nos 1 and 14 will carry 81 mm Mortars for smoke protection.

ASSAULT TRAINING CENTER
CONFERENCE
HQ ETOUSA

Annex IIIC to Field Order #1

NAVAL GUNFIRE SUPPORT
(For Initial Phase)

Time		Target	FSG	Ships	Remarks
From	To				
-90	-00	12 14	1	BB	12 - On notice that Ranger Bn has failed.
-90	-00	13 14	2	BB	13 - On notice that Ranger Bn has failed.
-20	-2	1	1 5	BB 4 DD	BB add occasional 14" to its regular secondary btry fires
-20	-2	2	6	2 DD	
-20	-2	5	2 7	BB 2 DD	
-20	-2	7	3	½ CL	
-20	-2	8	3	½ CL	
-20	-2	10	4	½ CL	
-20	-2	15	4	½ CL	
-2	-1	1		2 LCT(R)	

ASSAULT TRAINING CENTER
CONFERENCE
HQ ETOUSA

Annex IIID to Field Order #1

GRAPHICAL SCHEDULE OF NAVAL GUNFIRE SUPPORT

Support Group I (offshore)

BB-1 -------- Hypominus ------ Minutes ---------- Hypoilus ---
         60 55 50 45 40 35 30 25 20 15 10 5 H 5 10

14"

5"

BB-2
         60 55 50 45 40 35 30 25 20 15 10 5 H 5 10

Support Group II (offshore)

BB-3
         60 55 50 45 40 35 30 25 20 15 10 5 H 5 10

14"

5"

Support Group III (offshore)

ASSAULT TRAINING CENTER
CONFERENCE
HQ ETOUSA

Annex IV to Field Order #1

FIRE SUPPORT GROUPS

No.	Ships	Guns	Batteries	Support
1	1 BB	10 - 14" 8 - 5"	2) 4 2)	Division
2	1 BB	10 - 14" 8 - 5"	2) 4 2)	Division
3	1 CL	12 - 6"	2	Division
4	1 CL	12 - 6"	2	Division
5	4 DD	24 - 5"	8	Red Beach Force
6	2 DD	12 - 5"	4	Blue Beach Force
7	2 DD	12 - 5"	4	Yellow Beach Force

Time From	To	Target	FSG	Ships	Remarks
-2	-1	2		1 LCT(R)	
-2	-1	5		1 LCT(R)	
-2	-5	4	5	4 DD	
-2	-5	6	2 7	BB 2 DD	
Thereafter all FSG on call or on additional scheduled fires					

ASSAULT TRAINING CENTER
CONFERENCE
HQ ETOUSA

ANNEX VA TO FO I

METHOD OF APPROACH TO SPECIAL SITUATION

1. <u>Organization of landing beach.</u>

    a. Typical plan is shown for Woolacombe beach - (Similar layout can be adapted to Saunton Sands).

    b. <u>Initial Plan:</u> For beach exits shown on Sketch #1. This work to be performed by two engineer companies. (See Engr Plan Sec III).

    c. <u>Final Plan:</u> for beach exits and beach layout shown on sketch #2. This work to be completed by the shore party engineers.

2. <u>Organization of landing area back of beach.</u>

    a. Typical plan for Woolacombe area shown on sketch #3. (similar plan will apply to Braunton area).

    b. The organization of this area is based upon the following:

    (1) Initial reserves for assault regiments to be brought in with reserve battalion and assembled in Woolacombe and Braunton for distribution.

    (2) Initial assembly area for reserve battalion of assault regiments Woolacombe and Braunton.

    (3) Initial assembly areas in Woolacombe area for three combat teams of the follow up division shown on the sketch 3. (A similar layout will apply to the Braunton area). Vehicles and personnel will move out from boats as soon as they land directly to these initial assembly areas where they will be organized and moved out at once to secure their initial objectives. Woolacombe will be used as an intermediate vehicle port for <u>disabled vehicles</u>. Disabled vehicles will be towed off the beach immediately to the intermediate port for repair and cleared to their proper CT areas as soon as possible.

    (4) Assuming that the follow up division pushes on to secure final bridge head, ammunition, gas and oil, rations, and water, supply points will be established at Braunton and in area about 3000 yards due east of Woolacombe. To be established immediately after follow up division clears the proposed supply areas. Sketch #3 shows these supply points for the Woolacombe area.

3. <u>Engineer Plan</u>

    Woolacombe area (similar plan for Braunton area)

    a. Clear and mark two 16 yard passages thru beach minefield from water line to beach exit at X. (1 Company). Land immediately following assault infantry (probably H plus 30).

    b. Prepare road thru lanes cleared and maintain. (1 Company). Land at H plus 1:30. Will need about 500 yards landing mat bulldozer and picks and shovels.

c. Remove obstacles, repair, and maintain roads from beach exits forward. (1 Company). Land at H plus 2:00. Bulldozers other road equipment.

d. Prepare additional vehicle exit road 400 yards north of main exit, (1 company to commence after completing task c.)

e. Upon arrival of shore party engineer companies on a., b., d., turn over tasks and assemble in front of Woolacombe Hotel for further instructions.

f. Shore Party Engineers to widen gaps in obstacles and clear an area 200 by 1000 yards by D plus 4 and 200 by 2000 by D x 4. Meantime troops landing must be moved at once to initial assembly areas well back from the beach, there they are checked and reorganized to move inland.

ASSAULT TRAINING CENTER
CONFERENCE
HQ ETOUSA

Annex VB to Field Order #1

ORGANIZATION OF ASSAULT COMPANIES

Boat No	Personnel	Armament	Materiel or Explosives
1.	1 Officer (Inf)	Normal	
	16 Riflemen (1½ sqds)	Normal	
	2 Gunners	1-LMG 30 Cal.	
	3 Gunners	1-Mortar 60 mm	
	8 Engineers	Normal	
2,3, 4,5, 6,7.	Same as No 1, except 2 officers are Engrs and riflemen of No.7 have 3 BAR		

PERSONNEL			ARMAMENT
Inf	Officers - 5		Rifle, M-1 Cal.30 - 120
	EM - 161		Carbines, Cal.30 - 51
			Rifle, 03, Cal .30 - 9
Engr	Officers - 2		BAR, Cal.30 - 12
	EM - 49		Mortar, 60 mm - 7
	Total 217		

ASSAULT TRAINING CENTER
CONFERENCE
HQ ETOUSA

ANNEX VC TO FIELD ORDER I

PERSONAL EQUIPMENT LIST

The following is the equipment to be carried over the beaches by troops in the assault waves:

1. Basic Arms

2. _____ rds if armed with M-1 or Carbine (except ammunition bearers)

3. Helmet

4. Gas Mask

5. 1 Canteen of water

6. Pkg K Rations

7. 2. D. Rations

8. Bayonet where called for.

ASSAULT TRAINING CENTER
CONFERENCE
HQ ETOUSA

ANNEX VD TO FO 1

FAR SHORE DETACHMENT

	(Per Bn)	
	Off	EM
2 Roadbuilding Teams (2 Angle Dozers)	2	40
1 Obstacle Removing Team (1 6 x 6 W/Winch & 1 Crane)		14
1 Decontamination Team (1 Angle Dozer)		14
1 Signal Section (1 ¼ ton truck)	1	27
1 MP Detachment		10
	3	105

Reconnaissance Elements

Message Center

Naval Beach Party (Regulate boat traffic and salvage)

Beach Marking Section (Beach limits - roadways)

FAR SHORE GROUPS

3 Bns Engrs	1 Cml Decon Co
1 Med Bn Amph	1 Det Grp R Sta
1 Beach Sig	2 AA Bns AW S-M
1 QM Serv Bn	1 Bn Engrs (C)
4 Amph Truck Cos	1 TD Bn
1 MP Co	2 Ranger Bns
1 Ord Maint Co (M)	1 Prcht Regt
1 Ord Am Co	1 M Tk Bn

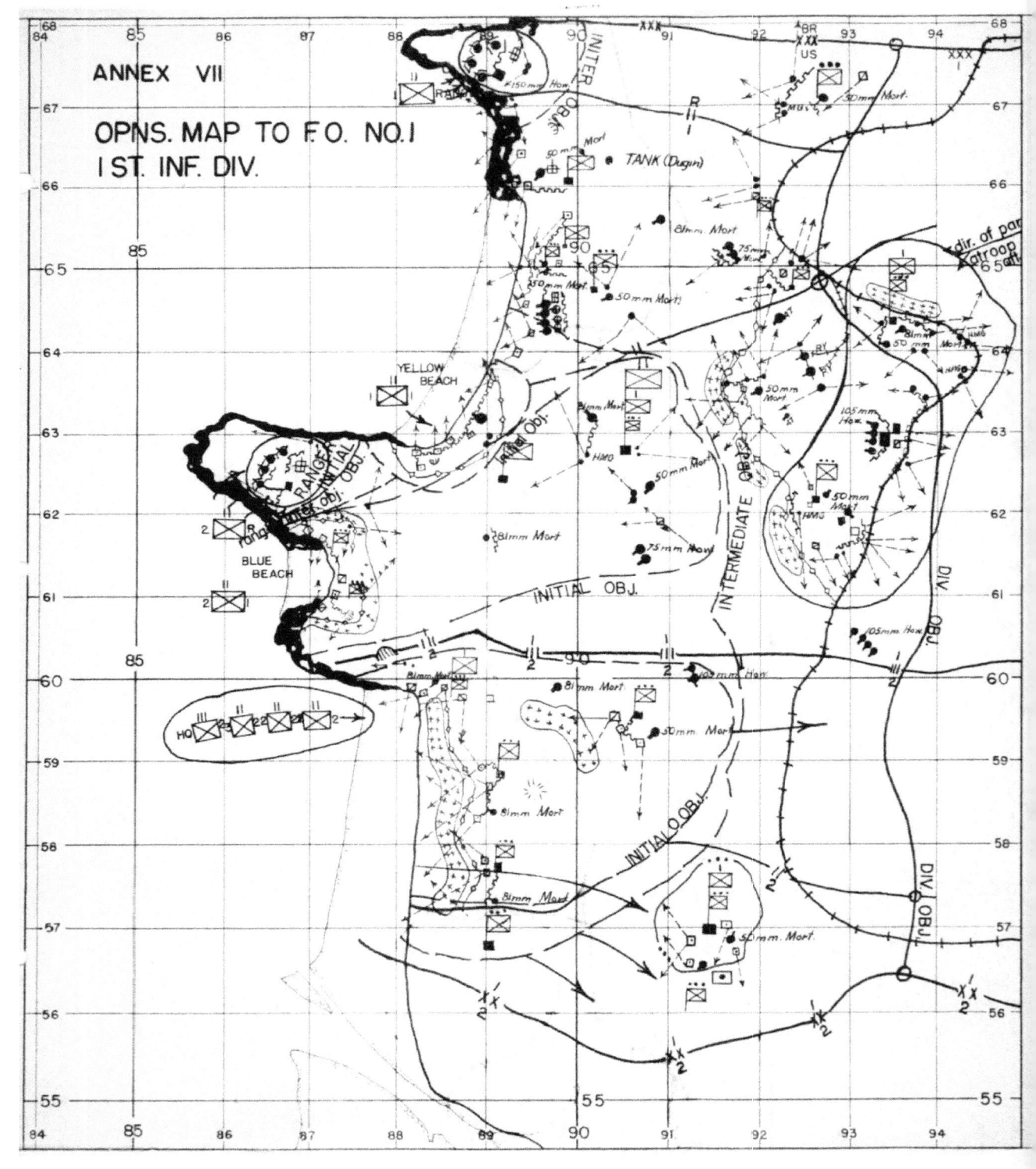

PHASE I
ANNEX VI A

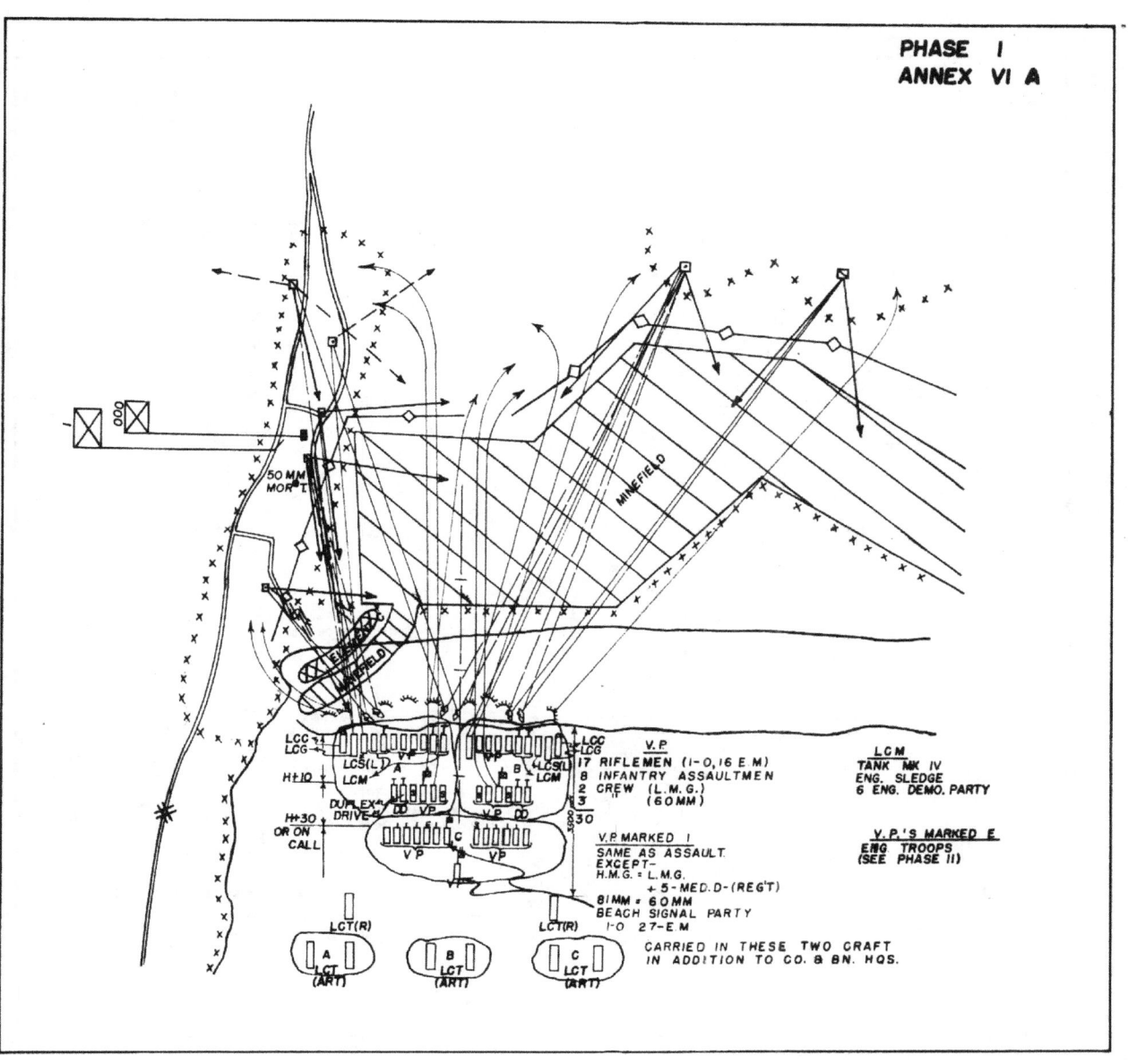

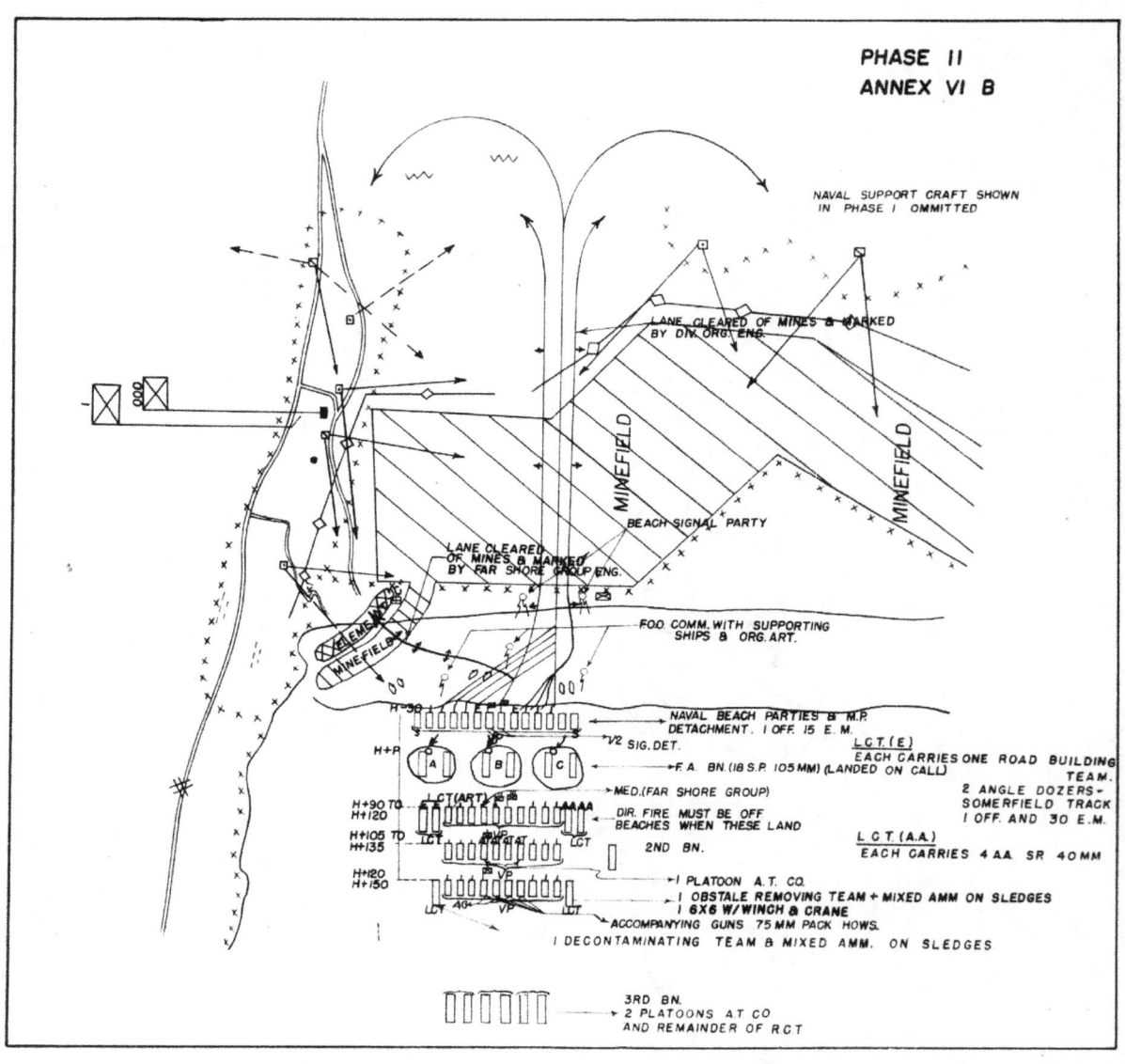

ANNEX VI C

NAVAL SUPPORT CRAFT
SHOWN IN PHASE I OMITTED.

1ST FOLLOW UP BN   2ND BN OBJECTIVE

ASSAULT BN OBJECTIVE

MINEFIELD

AA   AA

ELEMENT OF MINEFIELD

H+150
LCT   VP   LCT
H+

LCI(L) LST LCI(L)

PHASE III

3RD BN +
(1) LIGHT T.D. BN
(2) 2 PLATS A T CO
(3) ALL COMBAT VEHICLES (INF)
(4) 1 AA AW (S.P.) BTRY.
(5) 7 TANKS FROM TANK CO
(6) REMAINDER MED CO
(7) 1 FAR SHORE BN
(8) 6 AMM CARRIERS OF 105 AMM

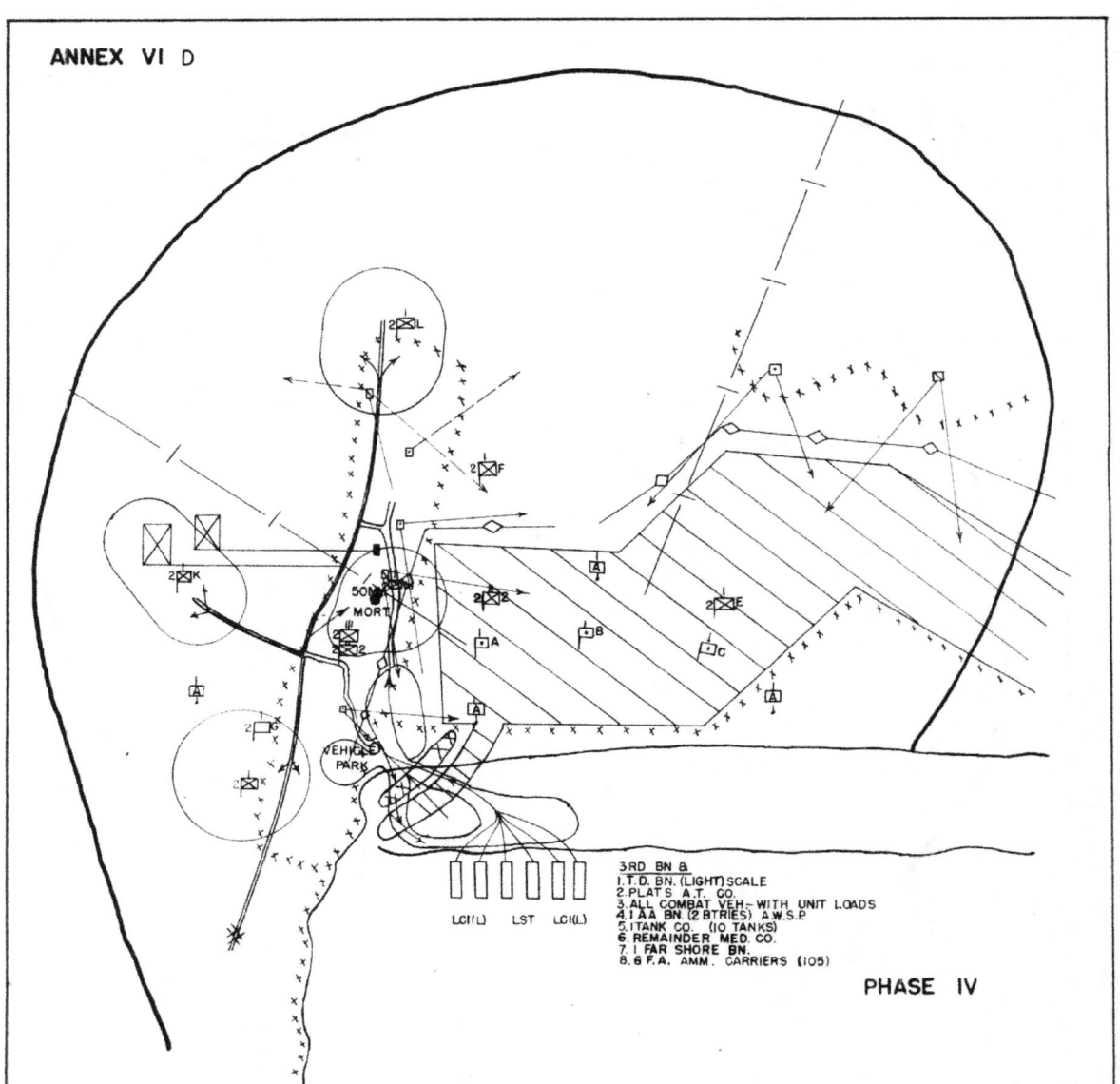

EXERCISE 3.

JR

ASSAULT TRAINING CENTER
CONFERENCE
HQ. ETOUSA

OPERATIONS
FIELD EXERCISE U.S.

Maps: Operations Map

I - INTRODUCTION

   1. Purpose

      a. To provide a tactical situation which requires a division to make a landing-assault under realistic conditions and to provide for, (or represent to), this unit the type of support from sea and air which it could logically expect.

   2. Methods of presentation

      a. A General Situation, (air, Navy and ground), is included to show only the necessary part of the Force Plan which must be known to insure a general picture of the type operation.

      b. A Special Situation "10th Div" is included to show in more detail the tactical plan of the division commander, orders that would be received from division, and support which could be furnished.

   3. Assumptions - (See Appendix to Annex # 3, map)

      a. The BARNSTAPLE BAY area is assumed to be part of the coast of FRANCE. The coast is assumed to be as shown on the map between MORTE POINT and HARTLAND POINT, (both incl), and to run due north from ROCKHAM BAY and due south from HARTLAND POINT.

      b. All territory to the east of this coast line is assumed to be on the continent of Europe.

II - GENERAL SITUATION

   1. Mission of Task Force

      a. Recent directives from GHQ Allied Forces established TASK FORCE U.S. Units of the force are now located in HOMELAND, approximately 50 miles southwest of LANDS END. Upon completion of training, Force U.S. will sail, date to be announced, with the mission of making a landing and establishing a beachhead on the NORTH DEVON shore between BULL POINT and APPLEDORE. TASK FORCE U.S. is suppoted by the 20th Air Force (Tactical), now based in HOMELAND. 8th Air Force and the Metropolitan Royal Air Force also based in HOMELAND, continue to wage a strategic offensive.

      b. The attack of TASK FORCE U.S. is accompanied by a smiliar attack on a beachhead area some 10 miles to the north by a British Force called FORCE GEORGE.

For purposes of this exercise it is assumed that the components of FORCE GEORGE, ground, air, and sea are the equivalent in strength of Force U.S.

2. Order of Battle

   a. Ground

      Hq Task Force U.S.
      Army Troops Incl 101st AA Brig
      XVI Corps
         10th Div
         11th Div
         12th Div
         Corps Troops
      XVII Corps
         13th Div
         14th Div
         15th Div
         Corps Troops
      20th Armored Div
      30th Airborne Div
      99th Paratroop Regt
      100th Ranger Regt

   b. Air

      Hq 20th Tactical Air Force
      9 Medium Bomb Groups (B-25's and B-26's)
      4 Light Bomb Groups (A-20's)
      6 Dive Bomb Groups (A-36's)
      13 Fighter Groups (7 gps P-47 6 gps P-51)
      1 Night Fighter Group
      9 Troop Carrier Groups (C-47 w/CG-4 Gliders)
      2 Observation Groups (P-51)
      1 Photo Group (F-4's and F-5's)

   c. Navy

      6 Battleships
      6 Cruisers
      24 Destroyers
      4 LCR
      16 LCG
      4 LCF
      organized into seven fire support groups.

3. Enemy Situation

   a. Ground: The enemy holds the entire coast in both directions from the BARNSTAPLE BAY area, in considerable strength. Divisions are holding sectors measuring 25 to 35 miles in width. All sectors are topographically similar to the BARNSTAPLE BAY sector, and are fortified on about the same scale. One Panzer division is believed to be now located in the vicinity of TAUNTON, and capable of closing in the BARNSTAPLE BAY area within 24 hours. Another Panzer division can be considered to be within 48 hours of the beachhead area.

   b. Air: The hostile Air Force is believed to have an available strength in SOUTHERN ENGLAND of not over 300 single engine fighter aircraft and 300 medium or light, (level or dive), bomber aircraft. Airdromes are as indicated on Maps Ordnance Survey of Great Britain - AIR - (¼ - 1 mi) plus two additional airdromes as shown on Appendix 1 to Annex #3.

4. **Extracts from Outline Plan, Task Force U.S.**
(See Map 2.)

    a. **Ground:** "Force U.S. will seize and establish a beachhead between MORTE POINT and BIDEFORD, both inclusive, see Map 1. Appendix 1 to Annex #3"

"Seize the ports of APPLEDORE and BARNSTAPLE"

"Occupy a defensive position (See Map 1 Appendix 1 to Annex #3) to permit our continued use of ports and airfields for the advance from beachhead line".

"For Corps and Division boundaries, initial and final objectives, see Map 2, Appendix 1 to Annex #3"

" XVI Corps will attack on column of divsions, seize beachhead line and prepare to advance to the northeast. 30th Airborne Division, 100th Ranger Regt (less 1 Bn) and 1st Bn 99th Paratroop Regt are attached to 16th Corps during the assault and until beachhead line is organized".

"XVII Corps, with 99th Paratroop Regt (Less 1 Bn) and 3rd Bn 100th Ranger Regt attached, attacks with two divisions abreast, captures CHIVENOR AIRDROME and BARNSTAPLE, seizes beachhead line and protects right flank of the Force".

"XVI Corps holds 11th Div in floating reserve to be committed only on orders from Force Hq Ship".

    b. **Air:** "20th Tactical Air Force supported by the 8th Bomber Command supports the assault by:

(1) Maintaining fighter cover over the transport area, and the approach and debarkation of assault waves on the beaches.

(2) Bombardment and ground attack missions <u>preparatory to the assault</u> with priority as follows:

    (a) Airdromes, (within enemy fighter range of beachhead area).

    (b) Beach defenses. (See Appendix 2 of Annex # 4, Map 1 "Plan of Supporting Fires)

(3) Bombardment ground attack missions <u>during the aasault</u> with priority as follows:

    (a) Movements of reserves.

    (b) Communication centers

    (c) Close support missions

    (d) Rear defense areas (See Appendix 2, to Annex # 4 Map #1 - "Plan of Supporting Fires")

(4) Smoke missions with priority to beach defenses on promontories dominating ABLE and BAKER BEACHES (See Appendix 1 to Annex #8 "Air Force Smoke Plan")

(5) Lift of Paratroop elements before and during the main assault and escort of glider borne elements."

"25 per cent of bombardment and ground attack aviation will be held in reserve for call direct from Task Force Joint Hq. Not over 50 per cent of total bombardment and ground attack aviation will be employed in prearranged missions immediately before and during the assault. Priority in close support pre-arranged missions to XVI Corps, up to 50 per cent of total sorties available".

"Air Support Parties with divisions and with combat teams of assault divisions. Air requests direct from Air Support Parties to Force Hq Ship".

"Arrival of first lift of glider borne elements of 30th Airborne Division will be timed to commence during last 15 minutes of preparatory air and naval bombardment and will be immediately preceded by attack of a large heavy bombardment formation of beach defenses of ABLE and BAKER BEACHES. Three groups of dive (fighter) bombers are reserved for escort and close support of Airborne Lifts".

"Following air units or detachments will cross the beahces with assaulting divisions:

* * * * * * *

Assault Division XVI Corps: 3 Air Support Parties
                             (3 Off - 15 EM)

                             2 Light Warning Sets
                             (2 Off - 24 EM)

                             4 Aircraft Warning Observers
                             (- 4 EM)

                             1 Weather Observing Section
                             (1 Off - 4 EM)

"For Bomb lines see Appendix (omitted)

"Tactical Reconnaissance sorties to prearranged points each 15 minutes over zones of action of each assaulting division. Brief radio reports broadcast to Force and Div Hq Ships and to Air Support Parties. Additional requests for reconnaissance to Control on Force Hq Ship".

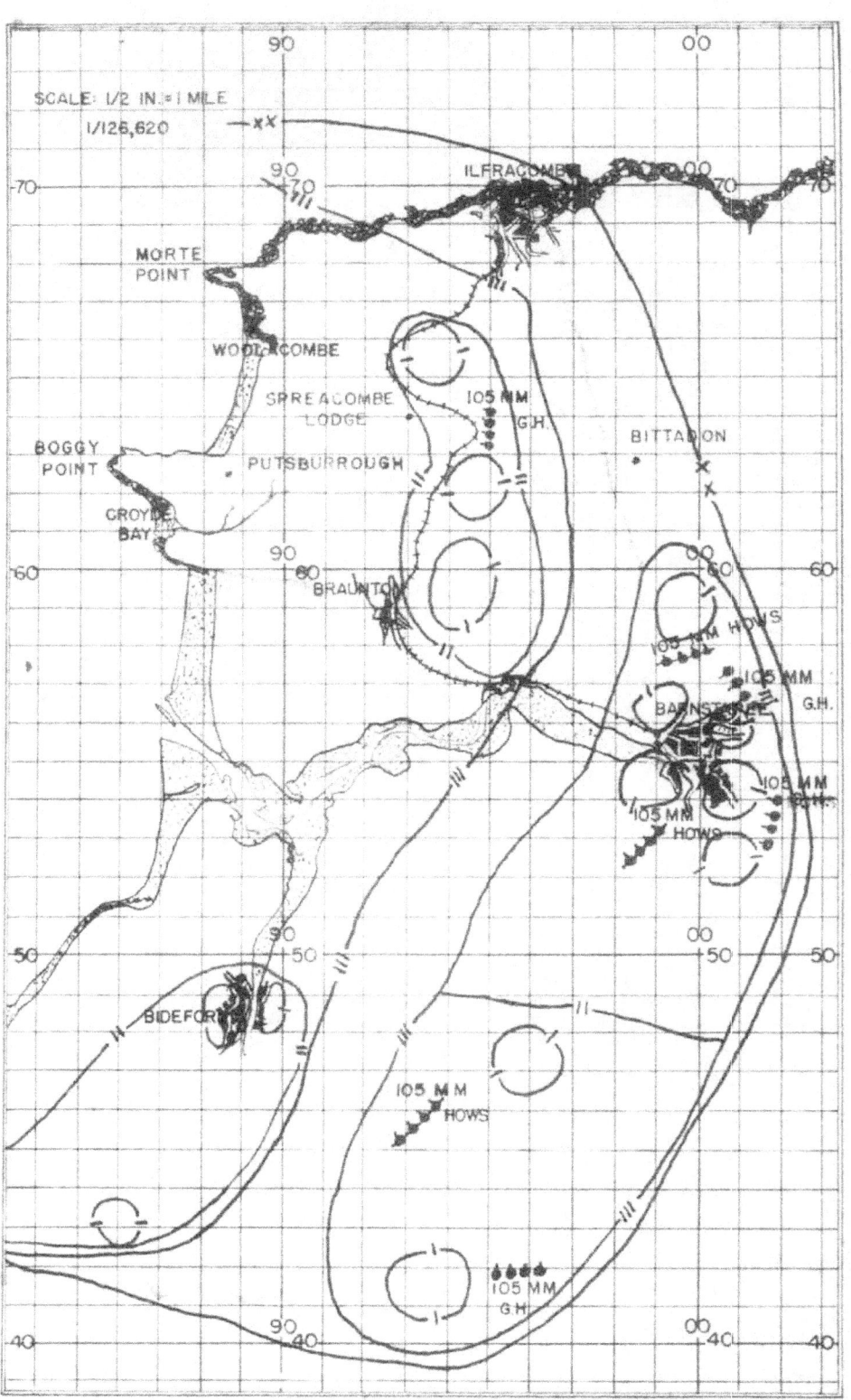

ASSAULT TRAINING CENTER
CONFERENCE
HQ. ETOUSA

PLAN
COMMANDING GENERAL 10th DIVISION (REINFORCED)

III Special Situation XVI Corps.

    CG XVI Corps assigns the 10th Inf. Div in the assault.

IV Special Situation 10th Inf Div (Reinf)

  1. a. Enemy dispositions: See map

    Enemy battalion support position vicinity SPREACOMBE LODGE is organized for two companies but believed garrisoned by only one. Two companies regimental reserve in our zone of action. Either may augment garrison in the support position or counter-attack to eject penetrations through the main defensive line.

    One Panzer division near EXETER available within 24 hours. An additional Panzer division can be expected within 48 hours.

    Two battalions of the Division reserve are located in the vicinity of BARNSTAPLE (See Annex #1)

  b. (1) XVII Corps on our right, boundary as shown on Opns Overlay Annex #3 Appendix 1 map #2. IV Corps (British) on our left.

    (2) Naval gunfire support by FSGs 1 to 7 inclusive and air support as shown in Annexes 4 and 6.

    (3) At H minus one hour and fifteen minutes the 99th Paratroop Regiment (Sep) will be dropped with the mission of attacking at H-15 to seize and hold the high ground in the vicinity of SPREACOMBE LODGE (See Annex #3 Appendix 1 Map #3.

    At the same time the 93rd Paratroop Regiment (30 A/B Div) will be dropped with the mission of seizing and holding the high ground west of BITTADON in order to prevent movement and counter-attacks of hostile reserves. Time of attack H-15.

    Beginning at H-15 the remainder of the 30th Airborne Division will be landed in the vicinity of BITTADON and occupy a defensive position (see Map # 3 Annex 3 Appendix 1) to prevent movement and counter-attacks by hostile reserves.

  2. a. This div:

    atchd:

    Combat group:

    100th Ranger Regiment (-1 Bn)

                901st Tank Bn (M)

                901st TD Bn

                AA Gp Hq   Note: AA Gp Hq and 2 Bns
                                        constitute 10th CA (AA)Gp
                501 AA Bn AW (SM)

                502 AA Bn AW (M)

                94th CML Bn (MTZ)

        Shore Party group:

                Engr Group Hq

                584th Engr Bn

                585th Engr Bn

                586th Engr Bn

                 81st Sig Co (Sp)

                281st Med Bn

                174th QM Bn (Serv)

                432nd QM Co (gas and oil)

                212th Ord Co (Am)

                766th QM Co (Rhd)

                767th QM Co (Rhd)

                    MP Co

will embark in ships and craft from Piers 1, 2, 3, and 4
HOMELAND Port of Embarkation on D minus 1 Day and on D-Day
force a landing in the Corps sector (see Opns Map #3
Annex #3, Appendix 1), seize and hold the indicated
objective. Main effort on the left.

      b. Details of boat movement - Annex #6 Appendix 3

      c. Lines of departure - Annex #6 Appendix 3

      d. D-Day and H-Hour to be announced.
           H-Hour is 60 minutes after nautical twilight
           D-Day is a day on which high tide is at H plus
two hours.

3. a. 28th Inf

        Attached: 28th FA Bn

                   Co A 10th Engr Bn

                   Co A 94th CML Bn (Mtz)

Co A 901st Tank Bn

Btry A 501st AA Bn AW

Co A 10th Med Bn

Far Shore Group

28th RCT	Far shore group contingent	
	Off.	EM.
2 roadbuilding teams (2 angle dozers)	1	26
1 obstacle removing team (6x6 w/winch & crane) truck mounted		14
1 decontaminating team (angle dozer)		14
1 Signal Sec (1 team - 1 jeep per team)	1	27
1 MP detachment		10
	2	91

Reconnaissance elements
Message center
Naval Beach Party -- (Regulate boat traffic and salvage)
Beach Marking Section - (Limits of beach. Roadways)

will seize BAKER BEACH, destroy enemy strong point northwest of PUTSBOROUGH and advance to Regt Int Beachhead Line in its zone of action.

    Formation: Bn landing teams in columns.

    b. 29th Inf

        Attached: 29th FA Bn

                 Co B 10th Engr Bn

                 Cos B & C 94th Cml Bn (mtz)

                 Co B 901st Tank Bn

                 Co B 10th Med Bn

        Btry B 502d AA Bn AW:

            29th Far Shore sub-group

29th RCT	Far shore group contingent	
	Off.	EM
4 road building teams (4 angle dozers)	4	52
2 obstacle removing teams (2-6x6 w/winch plus 2 crane)		56
2 decontaminating teams (2 angle dozers)		56
2 Signal Sec (2 jeeps)	2	54
2 MP detachments		20
	6	238

Reconnaissance elements
Message center
Naval Beach Party (regulates boat traffic~~ ~~)
Beach Marking Sec (beach limits - roadways)

will seize ABLE BEACHES GREEN AND YELLOW destroy enemy strongpoint at entrance to WOOLACOMBE corridor, and advance to Regt Int Beachhead line in its zone of action.

Formation: Two Bn landing teams abreast; one in floating reserve.

    <u>c</u>. Support Force:

        Comdg. Brig Gen _____ Asst Div Comdr

        Det Hq and Hq Co, 10th Inf Div

        Senior Beachmaster with detachment

        94th CML Bn (- 4 Cos)

        10th Inf Div Arty (-3 Bns)

        10th Rcn Tr (less dets)

        10th Sig Co (less dets)

        901st Tk Bn (less Cos A & B)

        901st TD Bn

        502nd AA Bn AW

        10th CA (AA) Gp

        10th Engr Bn (less dets)

will be prepared on two hours notice, on div order, to land any element on any of the beaches as directed; will establish div control of beachhead; to be prepared to employ elements to secure the Div Int Beachhead Line.

    <u>d</u>. Reserve:

        30th CT:

            30th Inf

            30th FA Bn

            Co C 10th Engr Bn

            Btry C 501st AA Bn AW

            Co D 94th Cml Bn (Mtz)

            Co C 10th Med Bn

in floating reserve will be prepared by F plus 2 hours, to move out from assembly area on div orders, and to land on any of the beaches as directed and attack in the zone of action of ~~either assault landing force~~.

_e_. 10th CA (AA) Gp will provide AA protection in the beach areas.

_f_. 100th Ranger Regiment (less 1 Bn)

First Bn 100th Ranger Regt will land at H minus 3 hours and seize and destroy the enemy battery and installations in MORTE POINT by H-65. Signal by 3 white parachute flares at H-65 will indicate success. After daylight success will be indicated by displaying a large red panel.

Second Bn 100th Ranger Regt will land at H minus 3 hours and seize and destroy the enemy battery and installations on BAGGY POINT by H-65. Signal by 3 red parachute flares at H-65 will indicate success. After daylight success will be indicated by displaying a large red panel.

_g_. 20th TAF and 94th Chemical Bn (mtz) dets attached assault CTs execute smoke missions in accordance with smoke plan (Annex #8 Appendix 2)

_h_. Other units to land on div orders.

_x_. (1) The prescribed success signals after landing will be promptly transmitted to the next higer unit.

(2) During the approach to the beach every precaution will be taken to eliminate unnecessary noise.

(3) All lights, except those essential for navigation and interior illumination are prohibited. All must be fully screened from observation by hostile forces.

(4) Unless it becomes necessary, in order to accomplish assigned missions, tactical unit commanders will not interfere with the operation or control of the landing craft.

(5) All units will leave liaison parties on the beach to meet and direct the next higher unit to land.

(6) Units of 94th Cml Bn revert to division control on div order.

(7) Landing craft will be prepared to release screening smoke without delay.

(8) Watches to be synchronized at 2300 hours D minus 1 day.

(9) Calls for direct Air Support through Air Support Parties with CTs.

(10) Calls for Naval gunfire through Shore Fire Control Parties with each assault LT.

(11) FA will revert to control of next higher artillery headquarters when that headquarters lands and establishes communication.

(12) CA AA units ~~will revert to div~~ control when CA (AA) Gp Hq lands.

(13) Shore Engr units will revert to control of next higher headquarters upon its establishment ashore.

(14) Radio silence until H-Hour, to be lifted earlier only on order of div commander.

(15) To lessen danger from our aircraft, front-line units will be prompt in displaying panels to warn friendly planes.

4. <u>a</u> Annex 2 Adm O 2

5. <u>a</u> Annex 7 Sig Opns

   <u>b</u> C.P's:

10th Inf Div	Hq Ship U.S.S._____
28th Inf	LSI_____
29th Inf	LSI_____
30th Inf	LSI_____
Div Arty	Hq Ship U.S.S._____
Engr Gp Hq	Hq Ship U.S.S._____

By command of Maj Gen A

OFFICIAL: C/S
     Y
     G-3

Annex 1 – Intelligence
      2 – Adm Order
      3 – Opn
      4 – Air Support Plan
      5 – Far Shore Group Plan (See Annex #2 Adm Order)
      6 – Naval operation
      7 – Signal communications
      8 – Smoke plan

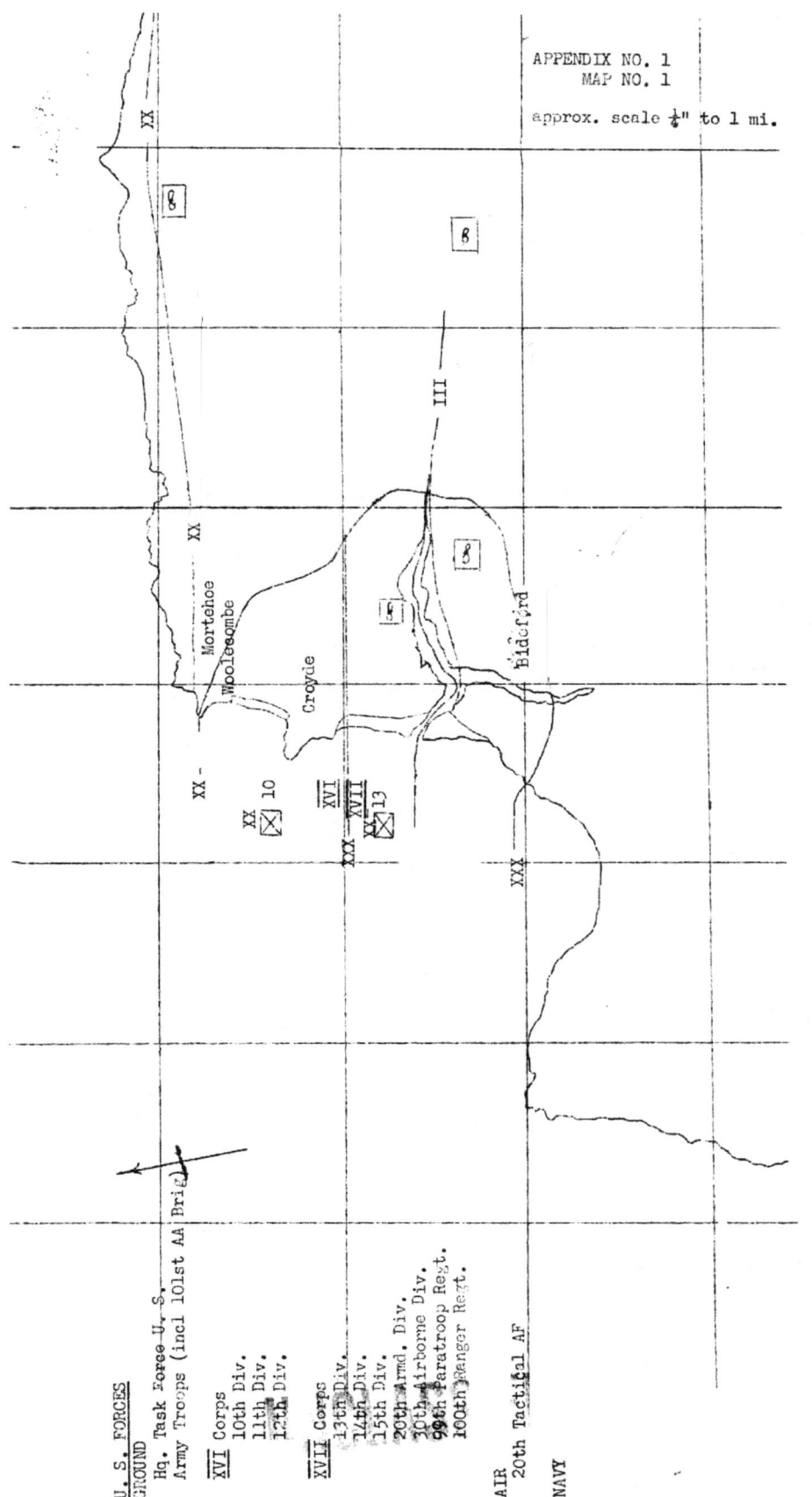

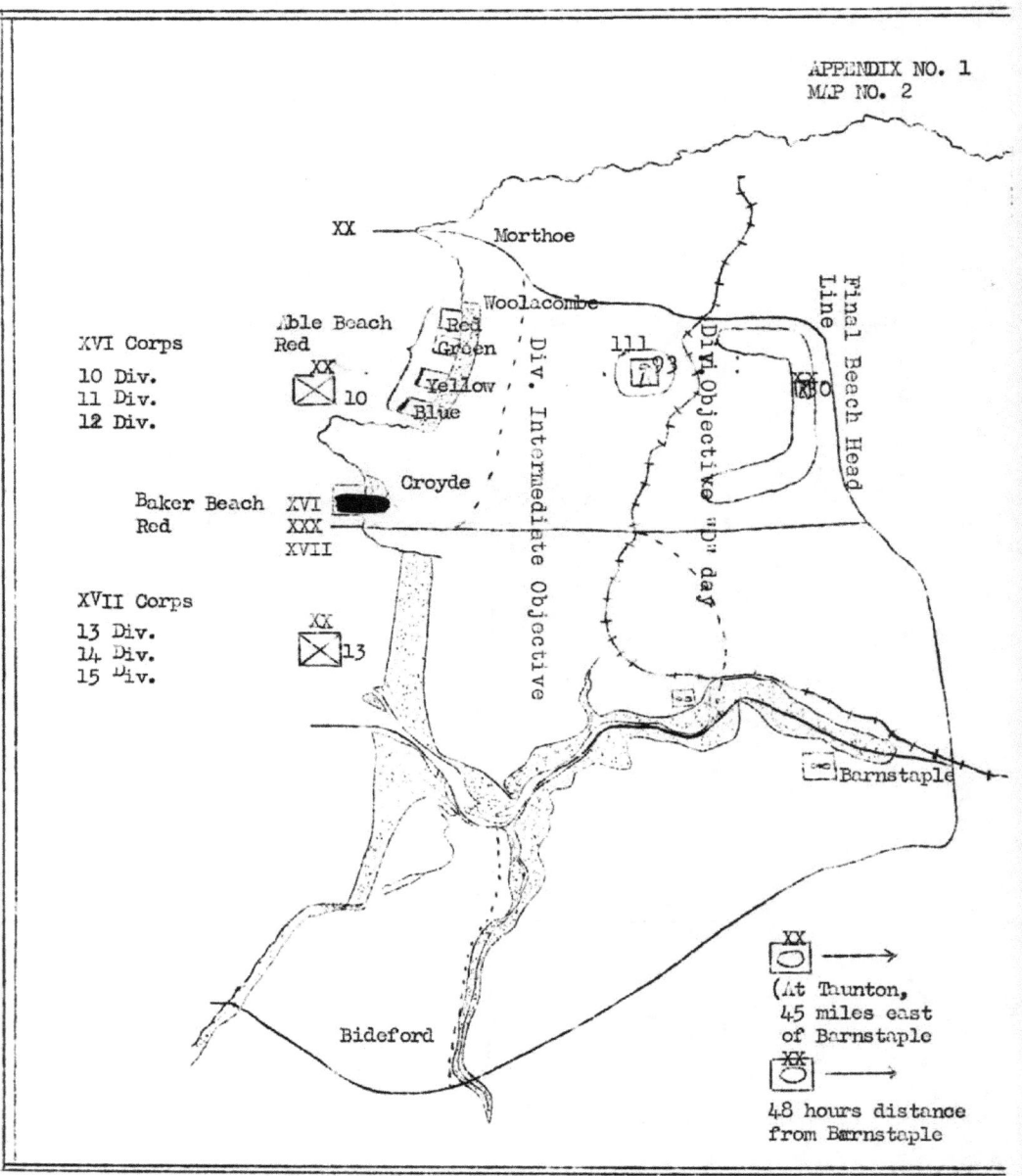

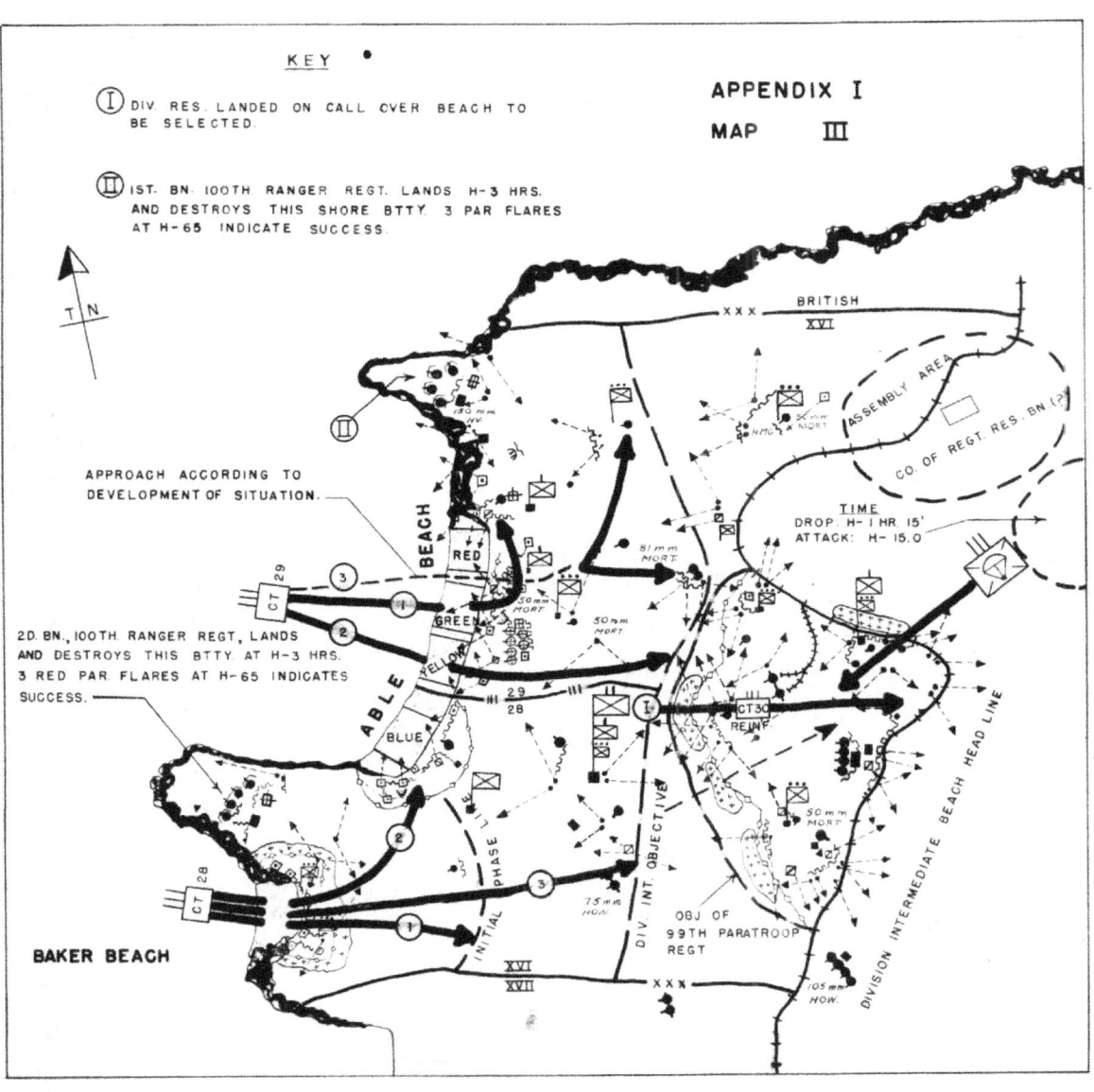

ASSAULT TRAINING CENTER
CONFERENCE

HQ ETOUSA

FIELD ORDER

FOR

COMBAT TEAM

Hq _____ CT

HOMELAND

_____, 1943

FO 1

Maps: Operations Map

1. a. Annex 1, Intelligence.

   b. (1) 10th Inf Div (reinf) attacks with 29th CT and 28th CT (28th CT on right) to seize beachhead area in MORTE POINT – point south of CROYDE BAY sector. Div will have naval and air support. 30th CT in Div res

   (2) XVII Corps on our right

   (3) IV Corps (British) on our left.

2. a. 29th CT,
   29th Inf
   29th FA Bn
   Co B 10th Engr Bn
   Co B 901st Tk Bn
   Co B 10th Med Bn
   Btry B 502d AA Bn AW
   Cos B and C, 94th Cml Bn (mtz)
   29th Far Shore sub-group

will debark from transports into landing craft during darkness early on D-Day; land at H-Hour, D-Day on ABLE BEACHES GREEN and YELLOW, destroy enemy strong point at entrance to WOOLACOMBE corridor, and advance to Regt Int. Beachhead Line in zone of action.

   b. Embarkation and Debarkation Table, Annex (omitted)

   c. Opns Overlay, Annex (omitted)

   d. Boat Allotment Table Annex (omitted)

   e. D-Day and H-Hour to be announced.

3. a. 1st LT:

        1st Bn 29th Inf
        Btry A 29th FA Bn
        1st Plat Co B 10th Engr Bn
        1st Plat Btry B 502d AA Bn AW
        Co B 94th Cml Bn (mtz)
        1st Plat AT Co
        1st Plat Cn Co
        1st Sect Far Shore Gp

will land at H-Hour D-Day on ABLE BEACH GREEN with co in assault; advance to cover beyond the dune line, change direction to the left, attack at H+40 and destroy enemy strong point at entrance to WOOLACOMBE corridor.

    b. 2nd LT

        2nd Bn 29th Inf
        Btry B 29th FA Bn
        2nd Plat Co B 10th Engr Bn
        2nd Plat Btry B 502nd AA Bn AW
        Co C 94th Cml Bn (mtz)
        2nd Plat AT Co
        2nd Plat Cannon Co
        2nd Sect Far Shore Gp

will land at H-Hour D-Day on ABLE BEACH YELLOW with 2 cos in assault, destroy enemy defenses in immediate front and advance to Regt Int. Beachhead Line (Open Overlay).

    c. Regt Hq 29th Inf
       29th FA Bn (less Btrys A, B, and C)
       Co B 10th Engr Bn (less 1st, 2nd and 3rd Plats)
       Btry B 502nd AA Bn AW (less 1st and 2nd Plats)
       94th Cml Bn (mtz) (less Cos A, B, and C)
       AT Co (less 1st, 2nd and 3rd Plats)
       Cn Co (less 1st, 2nd and 3rd Plats)
       Serv Co (less Dets)

embarked with Regtl Comd Gp, will be prepared to land on ABLE BEACH on order.

    d. Reserve:

        3rd Bn 29th Inf
        Btry C 29th FA Bn
        3rd Plat Co B 29th Engr Bn
        3rd Plat AT Co
        3rd Plat Cn Co

in floating reserve, will be prepared to land on ABLE BEACH on order.

    l. Air Support Party, embarked with Regtl Comd Gp, will land on order CO 29th CT

    x. (1) LT Comdrs will effect coordination with boat control officers on respective LSIs.

       (2) Success reports will be promptly made after landing.

       (3) Contact left to right, all units.

       (4) Naval gunfire support through Shore Fire Control Parties with assault landing gps after H+20.

(5) Every effort will be made to secure secrecy and surprise.

(6) No firing in assembly area or during approach to beach except to repel attack.

(7) Atchd units will revert to div control on order.

(8) Watches to be synchronized with official ship's time at 2300 hours, D minus 1 day.

(9) Full use will be made of Cml Cos for close support of assault.

(10) Troops constantly alert to danger of booby traps.

(11) To lessen danger from friendly aircraft, front-line troops will promptly display panels on call.

(12) Air support through Air Support Party at CT CP.

(13) Prompt report to be made if enemy makes use of toxic chemicals.

4. Annex 2, Adm O

5. a. (1) Annex _____, FO _____, 29th Inf Div Signal

   (2) Prescribed signals:

Meaning	Voice	Key	Lamp	Pyrotechnic
Distress Signal, boats	Apples	BDR	BDR	Signal, Grd White star parachute, M-17
Landing successful	Cherry	QTK	QTK	Signal, Grd Green Star Parachute, M-19
Lift Naval Gunfire	Quince	Nex	Nex	Signal, Grd Green Star Parachute, M-20
Display front line panels				Signal Aircraft Red Star Parachute, M-11

b. (1) CPs: 29th CT, aboard LSI _____
           1st LT   "   "   _____
           2nd LT   "   "   _____
           3rd LT   "   "   _____

After landing, LTs will leave liaison parties at beach to contact succeeding units.

c. Radio silence until H-Hour unless the movement is positively discovered. Breaking of radio silence earlier than H-Hour only on authority of Div Comdr.

By order of Colonel A

Ex O

OFFICIAL:
   Y
   S-3

Annexes:
   1. Intelligence
   2. Adm O
   3. Embarkation and Debarkation Table
   4. Opns Overlay
   5. Boat Lift Table
   6. Boat Assignment Table

- 3 -

ASSAULT TRAINING CENTER
CONFERNECE
HQ. ETOUSA

ADMINISTRATIVE ORDER # 2.   ANNEX # 2

1. Supply.

    a. Rations:

        (1) Supply points: Dumps initially selected, developed and marked by shore party of each R.C.T. on ABLE BEACH RED and BAKER BEACH.

        (2) Dump distribution to begin at H plus 48 hours.

        (3) Plan of supply,

            (a) Individual reserve - 2 K Rations carried by each man.

            (b) Initial reserve - 2 K Rations for entire force landed by H plus 48 hours.

            (c) Beach reserve - 4 C Rations for entire force established by H plus 96 hours and maintained at that level until ports are in operation.

    b. Ammunition:

        (1) Supply points: Initially engineer shore dumps on ABLE BEACH RED and BAKER BEACH.

        (2) Plan of supply.

            (a) Individual reserve: as prescribed by R.C.T.

            (b) Initial reserve: One unit of fire by H plus 24 hours plus one additional unit of fire by H plus 36 hours.

            (c) Beach reserve: After H plus 36, ammunition dumps will be built up to 6 units of fire, to be maintained at that level.

    c. Gas and Oil.

        (1) Distributing points: Shore dumps on ABLE BEACH RED and BAKER BEACH.

        (2) Plan of supply:

            (a) All vehicles will embark with gasoline tanks 95% full.

                (i) All vehicles will carry a reserve as prescribed by R.C.T.

            (b) 5,000 gallons regular, 92 octane and 100 octane by H plus 48 hours.

d. Water.

(1) Distributing points. Established by engineer shore units.

(2) All units down to and including squads will be provided with and trained in the use of emergency chlorination material.

(3) Plan of supply:

(a) Individual reserve - All troops will land with one full canteen.

(b) Initial reserve - Each landing craft, landing after H plus 12 hours, will carry five full five-gallon cans plainly marked for content.

(c) Beach reserve 5,000 gallons for each landing force by H plus 48 in addition to initial reserve.

(d) Shore party engineers will develop resources upon landing without delay.

2. Evacuation.

a. Casualties:

(1) Clearing station:

(a) Shore group Med Bn open on beaches by H plus 3 hours.

(b) Division Med Bn open when indicated by tactical situation.

(2) Collecting station.

(a) Shore group Med Bn open on beaches by H plus 2 hours.

(b) Div Coll Co attached to each R.C.T. to open when indicated by tactical situation.

b. Burial.

(1) Burial by organization.

(2) Div Cemetery. To be announced later.

c. Prisoners of War.

(1) Collecting points: ABLE BEACH RED and BAKER BEACH.

(a) Organizations will deliver prisoners of war to shore installations as indicated.

(2) Inclosure: ABLE BEACH YELLOW

(3) Shore units will forward prisoners of war to inclosure to be evacuated to base as soon as practicable.

3. Traffic.

    a. Circulation. Held to absolute minimum.

        (1) M.S.R. By R.C.T.'s - Division later.

        (2) Priorities, By R.C.T. - Division later.

    b. Control.

        (1) Routes between water's edge and shore dumps to be controlled by shore units.

        (2) Routes forward of shore units installations to be controlled initially by R.C.T.'s Div. Control to be effected upon landing of Division headquarters.

    c. Construction and maintenance of routes.

        (1) Shore party engineers to select, develop, maintain routes from water's edge to shore installations.

        (2) Division engineers to select, develop, maintain routes in the area forward of the shore installations.

4. Service troops and trains.

    a. Bivouacs. Location of Service troops later (By Division).

5. Personnel.

    a. Stragglers.

        (1) Straggler line - beach line.

        (2) Straggler collecting points: to be established by R.C.T. and reported to Div Hq as soon as possible.

        (3) Division M.P. platoon to take over when Division Hq lands.

    b. Surplus baggage. To be stored in embarkation area and transported to the far shore when the tactical situation permits.

    c. Mail: suspended until further notice.

    d. Furloughs and passes: Suspended until further notice.

    e. Strength reports: Daily after H plus 2 days as of 2400 to reach Div Hq by 0700.

    f. Replacements:

        (1) Normal requests with strength reports. Emergency requests by quickest communication available to forward echelon Div Hqs.

    g. Civilian population.

(1) Immediately upon occupying a populated village or town, the R.C.T. C.O. will be responsible that civil officials are instructed that all people will be frozen in their homes until further notice; that any civilian found outside their homes will be evacuated as prisoners of war.

6. Miscellaneous.

   <u>a</u>. Reports.

   (1) S-4 reports will be submitted daily with the strength returns.

FIRE PLAN
Co 901st Tk Bn (less 1 Plt)
in Support of Assault Cos.
1st Bn, 29th Inf.

Tank No.	Initial Target		Second Target		Alternate Target	
	No	Type	No	Type	No	Type
1	6	MG	6	MG	7	MG
2	8	MG	8	MG	6	MG
3	9	AT	7	MG	6/8	MG
4	9	AT	9	AT	8	MG
5	9	AT	9	AT	8	MG
6	2	MG	2	MG	3	MG
7	3	MG	3	MG	2	MG
8	4	AT	5	MG	2/3	MG
9	4	AT	4	AT	3	MG
10	4	AT	4	AT	5	MG

Cos. A & B, Assault Bns

  4 Platoons as follows:

        1 Rifle Squad - 12 EM

        1 section { 1-60mm Mortar Squad / 1-LMG Squad / 1-Sergeant } - - - - - - - - 11 EM

        1 section:
            2 Flame throwers - - - 4 EM
            2 Rocket Projectors - - 4 EM
            Pole charges - - - - 2 EM
            Bangalore Torpedos - 2 EM
            Grenades - - - - - - 2 EM
            N.C.O.'s - - - - - - 2 EM
            Total   16 EM   16 EM

    Total Platoon   39 EM 1-O.

    Total Company about 185 EM & 5 officers.

Co. C., Assault Bn - T/O Company.
Co. D., Assault Bn - T/O Company.

Order of Landing

1st Wave	1st & 2d Pls Co A	1 Pl Tks	H
2d "	3d & 4th Pls Co A	1st & 2d Pls Co B	H ∤ 12
3d "	3d & 4th Pls Co B	Shore Elements	H ∤ 18
4th "	1 Pl Tks	Cml. Co (-2 Pls)	H ∤ 60

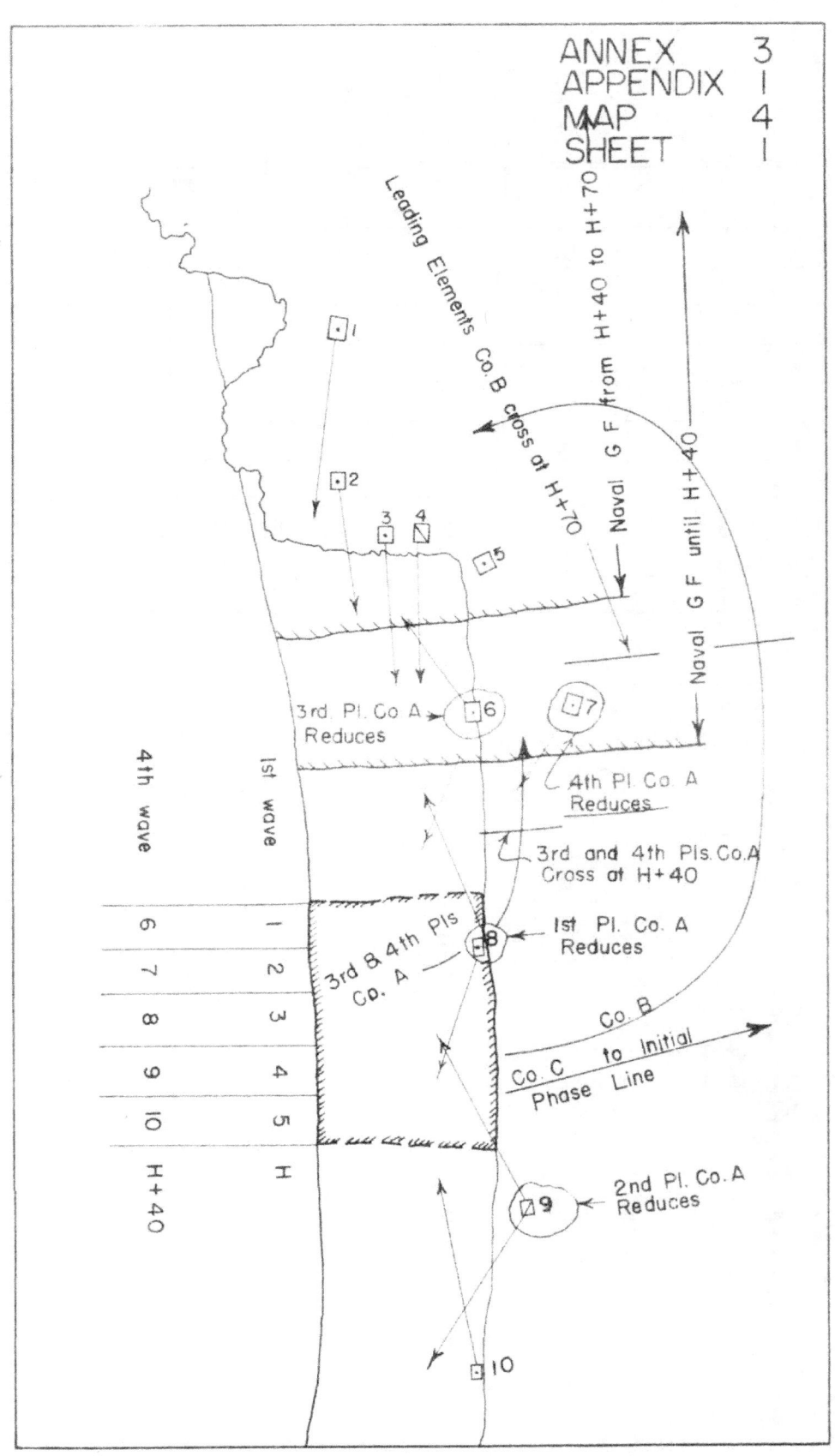

ASSAULT TRAINING CENTER
CONFERENCE
HQ. ETOUSA.

Annex #2
Appendix 2

ENGINEER PLAN

1. <u>Organization of landing beach</u>.

   <u>a</u>. Typical plan is shown for WOOLACOMBE BEACH – (Similar layout can be adapted to SAUNTON SANDS).

   <u>b</u>. <u>Initial Plan</u> : For beach exits shown on <u>Sketch #1</u>. This work to be performed by two engineer companies (See Engr Plan, Sec III).

   <u>c</u>. <u>Final Plan</u>: for beach exits and beach layout shown on <u>sketch #2</u>. This work to be completed by the shore party engineers.

2. <u>Organization of landing area back of beach</u>.

   <u>a</u>. Typical plan for WOOLACOMBE AREA shown on <u>sketch #3</u>. (Similar plan will apply to BRAUNTON AREA.)

   <u>b</u>. The organization of this area is based upon the following:

      (1) Initial reserves for assault regiments to be brought in with reserve battalion and assembled in WOOLACOMBE and BRAUNTON for distribution.

      (2) Initial assembly area for reserve battalion of assault regiments WOOLACOMBE and BRAUNTON.

      (3) Initial assembly areas in WOOLACOMBE area for three combat teams of the follow up division shown on the sketch 3. (A smilar layout will apply to the BRAUNTON area.) Vehicles and personnel will move out from boats as soon as they land directly to these initial assembly areas where they will be organized and moved out at once to secure their initial objectives. WOOLACOMBE will be used as an intermediate vehicle park for <u>disabled vehicles</u>. Disabled vehicles will be towed off the beach immediately to the intermediate park for repair and cleared to their proper CT areas as soon as possible.

      (4) Assuming that the follow up division pushes on to secure final bridge head, ammunition, gas and oil, rations and water, supply points will be established at BRAUNTON and in area about 3000 yards due east of WOOLACOMBE. To be established immediately after follow up division clears the proposed supply areas. Sketch #3 shows these supply points for the WOOLACOMBE area.

3. <u>Engineer Plan</u>

   WOOLACOMBE area (similar plan for BRAUNTON area)

<u>a</u>. Clear and mark two 16 yard passages through minefield from water line to beach exit (1 Company) Land immediately following assault infantry (probably H plus 30).

<u>b</u>. Prepare road through lanes cleared and maintain. (1 Company) Land at H plus 90. Will need about 500 yards landing mat, bulldozer and picks and shovels.

<u>c</u> Remove obstacles, repair, and maintain road from beach exits forward. (1 Company). Land at H plus 2 hours. Bulldozers and other road equipment.

<u>d</u>. Prepare additional vehicle exit road 400 yards north of main exit, (1 company to commence after completing task <u>b</u>.)

<u>e</u>. Upon arrival of Shore Party Engineer, Companies on <u>a</u>., <u>b</u>., and <u>d</u>. turn over tasks and assemble in front of WOOLACOMBE HOTEL for further instructions.

<u>f</u>. Shore Party Engineers to widen gaps in obstacles and clear an area 200 by 1000 yards by D plus 2, and 200 by 2000 yards by D plus 4. Meantime troops landing must be moved at once to initial assembly areas well back from the beach; there they are checked and reorganized to move inland.

SKETCH No. 1

BEACH TASKS FOR ENGINEERS WITH ASSAULT REGIMENT

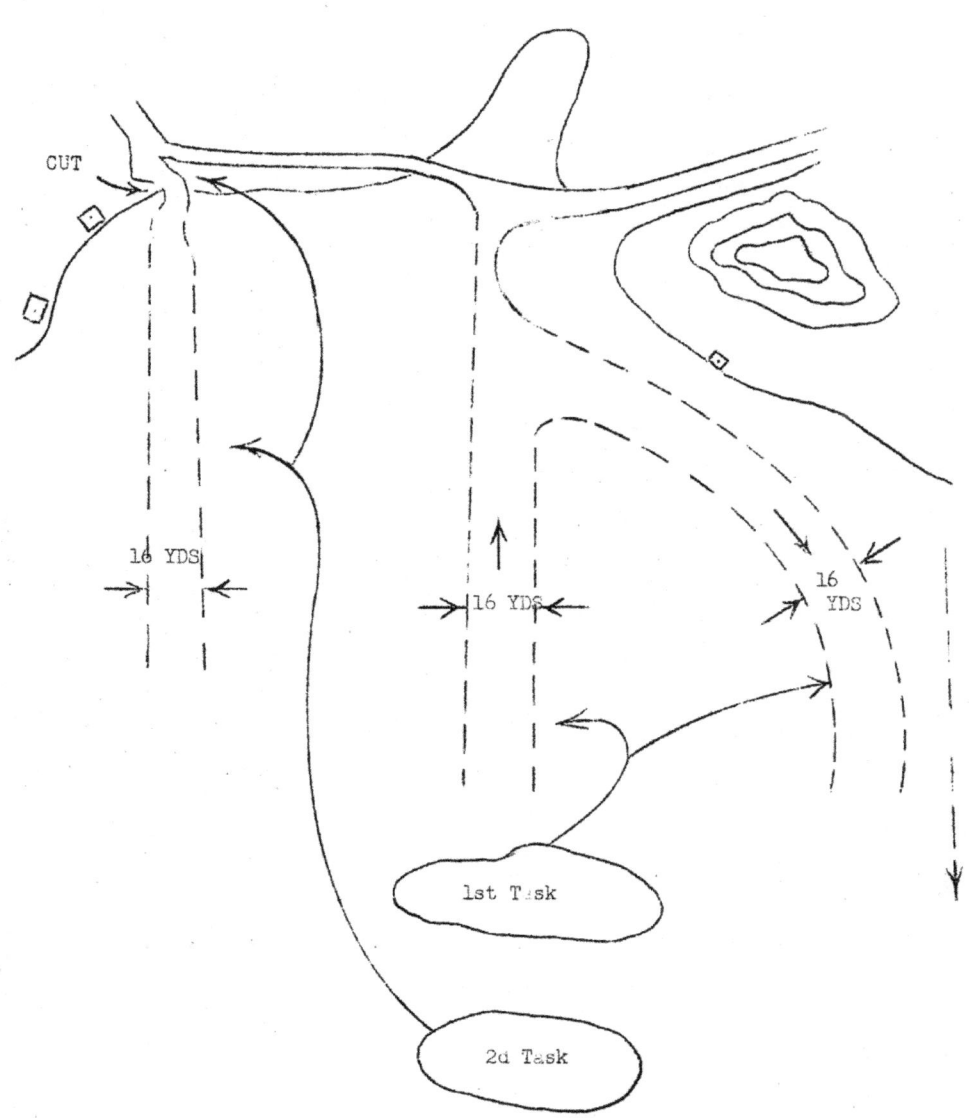

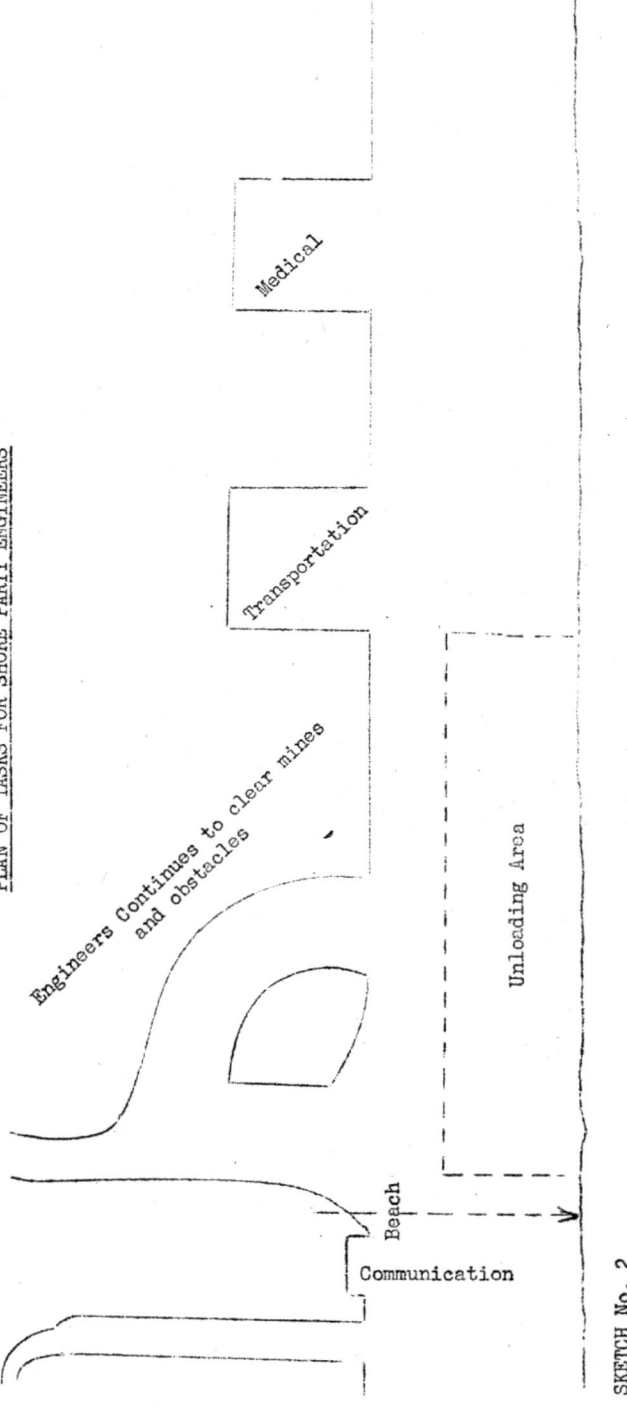

ORGANIZATION OF LANDING AREA BACK OF BEACH

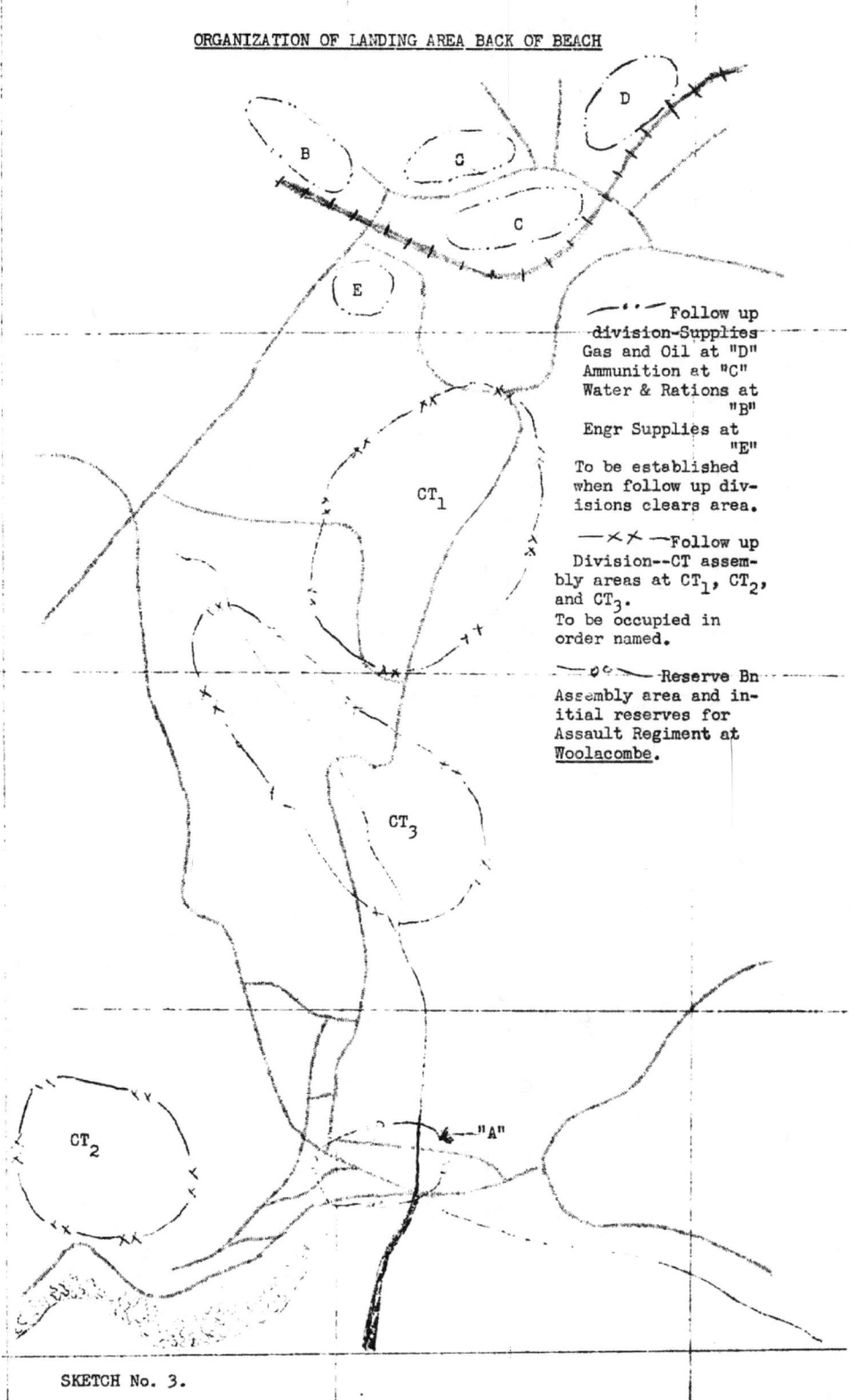

Follow up division—Supplies
Gas and Oil at "D"
Ammunition at "C"
Water & Rations at "B"
Engr Supplies at "E"
To be established when follow up divisions clears area.

Follow up Division—CT assembly areas at CT_1, CT_2, and CT_3.
To be occupied in order named.

Reserve Bn Assembly area and initial reserves for Assault Regiment at Woolacombe.

SKETCH No. 3.

ASSAULT TRAINING CENTER
CONFERENCE
HQ. ETOUSA

Annex #4
Appendix 1

THE AIR PLAN

The following are pertinent extracts from the Air Plan 20th Tactical Air Force that affect the operations and plans of the XVI Corps and its Assault Division Commander.

1. **Fighter Cover**

    a. 17 Fighter Groups are now available to the 20th Tactical Air Force. In addition, seven fighter groups will be made available from 8th Fighter Command for escort purposes and to reinforce the fighter cover.

    b. Fighter cover will be maintained on D-Day by squadron sorties on patrol over the landing operations, shipping, and area. These sorties are maintained on patrol of 30 minutes each; as follows: The following is a suggested number of sorties for the day upon which plans will be based. It will be changed as the results of the air battle dictate. It is estimated that each squadron can put an average of 12 aircraft over the target area.

PERIOD	AMOUNT OF FIGHTER COVER ON PATROL
H-60 to H-30	3 Squadrons
H-30 to H+30	8 Squadrons
H+30 to H+5 hours	6 Squadrons
H+5 hours to H+10 hours	4 Squadrons
H+10 hours to H+16 hours	2 Squadrons

Note that the figures here shown are based upon a squadron of 25 fighter aircraft making 3 to 4 twelve aircraft sorties during the day.

2. **Bombardment and Ground Attack Support for Assault of Task Force U.S.**

Narrative: a. During night prior to attack 8th Bomber Command bombed at night the following key communication centers viz: BARNSTAPLE, ILFRACOMBE, BIDEFORD, SOUTH MOLTON. This was a culmination of a four weeks period of strategic bombing during which the following towns were hit: BRISTOL, TAUNTON, EXETER, and PLYMOUTH. Also during this period, attacks by 65th Heavy Bomber Force and by medium Bombers of 20th Tactical Air Force were made on key airdromes affecting the assault or beachhead area.

-1-

<u>b</u>. For the assault, the Supreme Commander allotted 8 heavy Bomb Groups of 8th Bomber Command to support the air attack of 20th TAF on D-Day.

<u>c</u>. As a result of joint conferences by air and ground commanders, the general plan of the weight of air support mission to be allotted in the zone of action of each Corps was determined. The following table indicates the distribution of the weight of attack as it affected the XVI Corps. (See Chart "A" next Page)

3. Before proceeding to Schedules of Air Support Schedules the following Work Sheet is required. This work sheet is based upon the allottment of the 7 Groups shown above the prearranged missions in zone of action of XVI Corps. (The ½ group shown above for smoke, missions is included in Smoke Plan) The 7 Groups available must be distributed over a 4 hour period or from H-30 minutes to H+3½ hours inasmuch as no aircraft can repeat a sortie under four hours. For purpose of distribution, the 7 groups above are shown on the basis of the number of flights at 6 aircraft each they can make available. (See Chart "B" next Page)

A. DISTRIBUTION OF BOMBER AND GROUND ATTACK SUPPORT
   (With reference to XVI Corps)

Total Groups available to 20th TAF for Assault	No. of Groups employed on D-day to continue neutralization of airdromes	Other Task Force Missions	Force Reserve	Available for pre-arranged close support missions for Force	Available for air request from force as a whole	Groups available for pre-arranged missions in zone of action of XVI Corps
8 hvy Bomb Gps	-	1	1	6	-	4 Gps
9 med Bomb Gps	3	-	2	2	2	1
4 L Bomb Gps	-	-	1	2½	-	1 Gp for Bombing, ½ Gp for Smoke
6 Ftr Bomb Gps (A-36)	-	3 (For escort of Carrier Lift)	1	2	1	1
TOTALS 27 Gps	3	4	5	12½	3	7½

B. MASTER WORK SHEET - AIR SUPPORT IN ZONE OF ACTION XVI CORPS
   (No. of 6 aircraft attacks - 1st 4 hour period)

Time Interval (in minutes)	Tactical Air Force				From Bomber Command & from Ftr Bomber Escort of Airborne Lift		Grand Total (6 a/c sorties)
	Medium Flights (6 a/c each)	Light Flights (6 a/c each)	FI (D) (6 a/c each)	Totals	Bomber Command	3 Dive Gps or Airborn Cmd	
H-30 to H	3	3	4	10	32 (4 atks)	4 (1 atk)	46
H to H+30	1	2	2	5	0	0	5
H+30 to H+90	2	1	2	5	0	0	5
H+90 to H+150	1	1	2	4	0	0	4
H+150 to H+210	1	1	2	4	0	0	4
TOTALS	8	8	12	28	32	4	64

ASSAULT TRAINING CENTER
CONFERENCE
ACA HOUSE

AIR SUPPORT FOR PREPARATION (H-30 to H)

TIME	FLIGHTS 6 a/c ea	TYPE a/c	BOMBS or AM	APPROX TOTAL BOMBS	OBJECTIVE	MISSION	ALTERNATE OBJECTIVE OR MISSION
H-30	8	B-17	1000#	240 - 1000# 120 tons	2 and 3 (Woolacombe)	Neutralization communications in WOOLACOMBE.	AA No. L(North Btry) (Change VHF Signal before H-35)
H-30	8	B-17	1000#	240 - 1000# 120 tons	7 and 8	Neutralization of beach defenses.	No. 9 (South Btry) Change VHF signal before H-35)
H-30	1	B-25	500#	24 - 500# 6 tons	17	Neutralization of AA Btry	No. 1 - on VHF signal by H-25
H-30	8	B-17	1000#	240 - 1000# 120 tons	10 and 11 (Croyde)	Neutralization communications in CROYDE.	No. 1 - on VHF signal by H-30
H-25	1	A-36	500#	12 - 500# 3 tons	30	Neutralization of R Btry.	No. 9 on VHF signal by H-30
H-25	1	A-36	500#	12 - 500# 3 tons	34	Neutralization of 105 mm Btry.	
H-25	1	B-25	500#	24 - 500# 6 tons	17	Continue neutralization of AA Btry.	
H-20	1	B-25	500#	24 - 500# 6 tons	17	Neutralization of AA Btry.	
H-15	8	B-17	1000#	240 - 1000# 120 tons	2 and 3 (Woolacombe)	Neutralization Diversion for Airborne Lift.	No. 9 - on VHF signal by H-20

-4-

TIME	FLIGHTS 6 a/c ea.	TYPE a/c	BOMBS or AM	APPROX TOTAL BOMBS	OBJECTIVE	MISSION	ALTERNATE OBJECTIVE or MISSION
H-15	4	A-36	Clusters of 6 frag bombs (50 cal MG)	4800# frag bombs machine gun	31,34,35	Protection of glider landing	Targets of opportunity just west of SOUTHERN RY.
H-11	1	A-20	100# and clusters of 6 frag	16 - 100# 5 tons	2, 3, & 4.	Attack north end of ABLE Beach	
H-11	1	A-20	100# and clusters of 6 frag	16 - 100# 5 tons	6, 7, & 8	Attack south end of ABLE Beach	
H-11	1	A-20	100# and clusters of 6 frag	16 - 100# 5 tons	10 & 11	Attack CROYDE Beach	
H-5	1	A-36	100# and clusters of 6 frag	1200 pounds frag plus machine gun	2, 3, & 4	Ground strafe north end of ABLE Beach.	If obscured by smoke attack targets of opportunity between shore and SOUTHERN RY.
H-5	1	A-36	100# and clusters of 6 frag.	1200 pounds frag plus machine gun	10 & 11	Ground strafe south end of ABLE Beach.	If obscured by smoke attack targets of opportunity between shore and SOUTHERN RY.

BOMB LINES from H-30 to H   East of Shore line and West of SOUTHERN RAILWAY

AIR SUPPORT
(H to H-3½ hours)

TIME	FLIGHTS 6 a/c ea	TYPE a/c	BOMBS or AM	APPROX TOTAL BOMBS	OBJECTIVE	MISSION	ALTERNATIVE OBJECTIVE or MISSION
H	1	B-25	500# 100#	12 500# Bombs 6 100# Bombs 6 tons	30	Neutralization of Ry Btry	No. 17 - (AA Btry)
H+5	1	A-36	Frag Bombs MG's	1200 pounds frags	31, 34, 35	Neutralization	Targets of opportunity west of SOUTHERN RY.
H+15	1	A-36	Frag Bombs MG's	1200 pounds frags	28, 27, 26	Neutralization	Targets of opportunity in areas 24 and 22.
H+30	1	A-20	100#	96 - 100# bombs approx 5 tons	Air alert 5 m north of transport area for call direct fr ASP W/29th Inf		If no request by H+50 attack 24 & 22
H+30	1	A-20	100#	96 - 100# bombs approx 5 tons	Air alert 5 mi south of transport area for call direct fr ASP W/28th Inf		If no request by H+50 attack 14 & 15
H+45	1	B-25	500# 100#	12 500# bombs 60 100# bombs 6 tons	21 and 22	Neutralization	31, 34 and 35 if requested by ASP's with AB Div
H+45	1	B-25	500# 100#	12 500# bombs 60 100# bombs 6 tons	Communications Center at CROYDE	Neutralization	17 or 30, as directed by Air Support Control.

TIME	FLIGHTS 6 a/c ea	TYPE a/c	BOMBS or AM	APPROX TOTAL BOMBS	OBJECTIVE	MISSION	ALTERNATIVE OBJECTIVE or MISSION
H+60	1	A-20	100# bombs	96 - 100# bombs approx 5 tons	Air alert 5 mi north of transport area for call direct fr ASP W/15th Div		If no request by H+90 attack 26 & 24
H+75	1	A-36	Frag Bombs MG's	1200 pounds frags	14 and 15	Neutralization	Targets of opportunity on or east of route B-3230.
H+90	1	A-36	Frag Bombs MG's	1200 pounds frags	Air alert 5 mi south of transport area for call direct fr ASP W/30th AB Div.		If no request by H+110 attack targets of opportunity east of route B-3230.
H+105	1	A-36	Frag Bombs MG's	1200 pounds frags	21 and 22	Neutralization	Targets of opportunity on or east of route B-3230
H+105	1	A-36	Frag Bombs MG's	1200 pounds frags	14 and 15	Neutralization	27 and 26
H+125	1	B-25	500# 100#	12-500# bombs 60-100# bombs 6 tons	24 and 26	Neutralization	Targets of opportunity on or east of route B-3230
H+125	1	A-20	100# bomb	96 - 100# bombs approx 5 tons	Air alert 5 miles north of Transport area for call direct fr ASP W/10th Div		If no request by H+150 attack targets of opportunity east of B-3230

H+150 to
H+210   No planned missions beyond H+150 but following sorties can be made available to
XVI Corps during period H+150 to H+210. These sorties are:

        12 - A-36 aircraft
         6 - A-20 aircraft
         6 - B-25 aircraft

Corps or Div Comdr must submit to Air Support Control by H+90 the missions desired for these sorties to accord with troops' progress.

NOTE:

1. After H+3½ hours, sorties flown during period H-30 to H become available again to the Air Commander. These will be allotted and planned, or air request missions will be directed to accord with the situation of our own troops and information of the enemy that is known at that time. From this point on, more missions than formerly will be flown as a result of Air Support Requests, from Parties with assaulting divisions. Furthermore, after H+3½ hours the movements of reserves will assume first priority for air attacks,

-8-

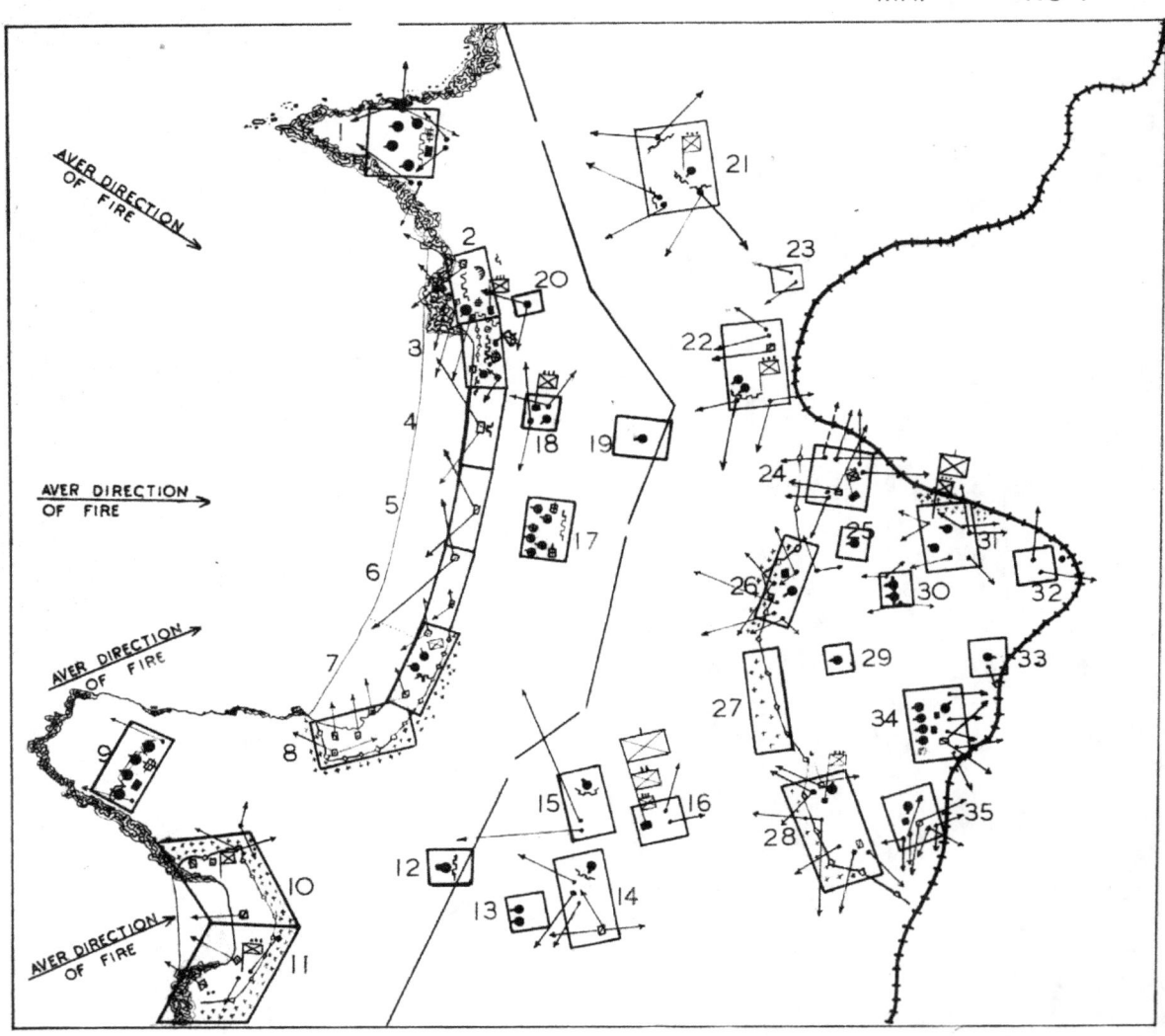

ANNEX   NO 4
APPENDIX NO 2
MAP   NO 1

ASSAULT TRAINING CENTER
CONFERENCE
HQ. ETOUSA

Annex #5

Far Shore Group Plan is included in Administrative Order #2 Annex #2.

ASSAULT TRAINING CENTER
CONFERENCE
HQ. ETOUSA

Annex #6
Appendix 1

GENERAL MISSIONS OF NAVAL TASK GROUPS FOR LANDING OPERATIONS WITH ARMY

1. <u>Naval task groups will</u>:

    1. (a) Provide adequte reconnaissance.

    (b) Provide protection against enemy naval forces during the landing operations.

    (c) Provide, man, equip and operate the small craft required for the operation and land personnel and material of the landing force, in accordance with the approved plan of the landing.

    (d) Support the operation by gunfire, aircraft, and screening operations from boat guns, mine sweeping and removing underwater obstacles.

    (e) Provide signal communications between ships and shore.

II. <u>Naval gunfire support</u> for 1 Division front of 4000 to 6000 yards.

    Assume: 2BB
               2CL
               8DD

III. <u>Preliminary Preparations for Naval Gunfire Support</u>

    1. Make out work tables based on number of 100 yard squares with density expressed in terms of 75mm per minute.

    2. <u>Operations Map showing</u> (From Army):

    (a) Landing beaches to be used.

    (b) Distribution of troop units for the landing.

    (c) Boundaries between units.

    (d) Main effort.

    (e) Objectives, and times in reference to H-Hour that troops are to capture each objective.

    (f) Dotted lines at convenient intervals to show the probable rate of advance of troops from the beaches to the <u>limit</u> of scheduled gunfire support.

    (g) Probable average direction of fire from supporting ships.

    (h) The zones which are to receive the most intensive fire support, that is, the target areas previously estimated to contain the most dangerous enemy resistance.

REQUIREMENTS FOR LANDING SHIPS AND CRAFT FOR LANDING I-DIVISION

Type	Speed	Assault Regiments (2)		Reserve Regiment (1)	Total (Division)	
		Assault Bns (4)	Reserve Bns (2)			
APA	15	1			4	APA
AKA	15	1 ?			(1)?	AKA
LCVP	10	(30)			120	LCVP
LCM(3)	10	(3)			12	LCMB
LCM(3)	10	(8)?			(8)?	LCMB
LCS(3)	16	(2)			8	LCS(3)
LST	11		4	12	20	LST
LCI(L)	15		4	12	20	LCI(L)
LCT(5)	10	5	4	12	40	LCT(5)
LCT(R)	10	1			4	LCR
LCG	10	4			16	LCG
LCF	10	1			4	LCF
LCC	15	1			4	LCC
LCS(L)	11	2			8	LCS(L)

Assault Regiments - 2 - 1 <u>Reserve</u>
Assault Bn   -4 - 5 Reserves
Bracketed craft are shipborne

FIRE SUPPORT GROUPS

No.	Ships	Guns	Batteries	Support
1	1BB	10-14" 8-5"	2) 2) 4	Divisions
2	1BB	10-14" 8-5"	2) 2) 4	Divisions
3	1CL	12-6"	2	Divisions
4	1CL	12-6"	2	Divisions
5	4DD	24-5"	8	(Red) ? Beach Forces
6	2DD	12-5"	4	(Blue) ? Beach Forces
7	2DD	12-5"	4	(Green) ? Beach Forces

Type	Speed	Yds/Min
APA	15	500
AKA	15	500
LCVP	10	333
LCM(3)	10	333
LCS(5)	16	533
LST	11	363
LCI(L)	15	500
LCT(5)	10	333
LCT(R)	10	333
LCG	10	333
LCF	10	333
LCC	15	500
LCS(L)	11	363

NAVAL GUNFIRE SUPPORT

Ship	Guns Broadside	Controls	Ammunition available Rounds: Type	Total rounds per minute	Estimated effect in 75mm per one round	Estimated total effect in 75mm per minute less 10%	Number of 100 yard squares with density $\frac{16}{\text{per minute}}$ 75mm : $\frac{12}{\text{per minute}}$ 75mm
1.BB	10-14"	2	350 : HC	15	16.0	216	14 : 18
	10-14"	2	150 : AP	15	6.0	81	5 : 7
	8-5"	2	2680 : AA	48	2.0	86	5 : 7
							24 : 32
L. CL	12-6"	2	1300: HC	60	3.7	200	12 : 17
1.DD	6-5"	2	600: AA HC	42	2.0	76	4 : 6

Notes: Assume the naval gunfire support for 1 Division front to be: (2.BB The equivalent total available 75mm density
(2.CL of fire
(8.DD

2.BB   48 : 64
2.CL   24 : 34
8-DD   32 : 48
Total   104 : 146

ASSAULT TRAINING CENTER
CONFERENCE

HQ ETOUSA

Annex #6
Appendix #2

COMMITTEE NO. 3

NAVAL GUNFIRE SUPPORT

Time From	To	Target	Rounds	FSG	Ships	Remarks
H-60	H-50	31	30	1	1BB	HC14"
H-60	H-50	34	30	2	1BB	HC14"
H-60	H-50	35	30	1	1BB	HC14"
H-60	H-50	28	30	2	1BB	HC14"
H-60	H-50	30	40	1-2	2BB	HC14" (20 rounds each group)
H-60	H-50	17	30	3	1CL	HC6"
H-60	H-50	1	--	1-2	2BB	HC14" On signal, assuming failure of rangers, shift from target 30
H-60	H-50	9	--	1-2	2BB	HC14" On signal, assuming failure of rangers, shift from target 30
H-50	H-40	2	50	3	½CL	HC6"
H-50	H-40	3	50	3	½CL	HC6"
H-50	H-40	4	50	4	½CL	HC6"
H-50	H-40	5	50	4	½CL	HC6"
H-50	H-40	6	50	2	1BB	20AP) 30HC) 14"
H-50	H-40	7	50	2	1BB	20AP) 30HC) 14"
H-50	H-40	8	50	2	1BB	20AP) 30HC) 14"
H-50	H-40	10	50	1	1BB	20AP) 14" 30HC)
H-50	H-40	11	50	1	1BB	20AP) 14" 30HC)
H-40	H-30	30	40	3	½CL	HC6"
H-40	H-30	31	40	4	½CL	HC6"
H-40	H-30	34	40	3	½CL	HC6"
H-40	H-30	35	40	4	½CL	HC6"
H-40	H-30	28	40	6	2DD	HC or AA 5"

- 1 -

Time From	To	Target	Rounds	FSG		Remarks
H-40	H-30	17	40	7	2DD	HC or AA 5"
H-40	H-30	2	40	2	1BB	10AP) 30HC) 14"
H-40	H-30	3	40	2	1BB	10AP) 30HC) 14"
H-40	H-30	7	40	1	1BB	10AP) 30HC) 14"
H-40	H-30	8	40	1	1BB	10AP) 30HC) 14"
H-30	H-15	4	60	3	1CL	6" HC
H-30	H-15	5	60	4	1CL	6" HC
H-30	H-15	6	60	5	2DD	5" HC
H-30	H-15	17	60	5	2DD	5" HC or AA
H-30	H-15	30	60	1	1BB	5" HC
H-30	H-15	31	60	2	1BB	5" HC
H-30	H-15	34	60	1	1BB	5" HC
H-30	H-15	35	60	2	1BB	5" HC
H-30	H-15	28	60	1	1BB	5" HC
H-15	H-2	4	60	5	4DD	5" HC or AA
H-15	H-2	5	60	5	4DD	5" HC or AA
H-15	H-2	6	60	6	2DD	5" HC or AA
H-15	H-2	17	60	3	1CL	6" HC
H-15	H-2	10	60	4	½CL	6" HC
H-15	H-2	11	60	4	½CL	6" HC
H-15	H∓70	2	90	7	2DD	5" HC or AA
H-15	H∓40	3	60	7	2DD	5" HC or AA
H-15	H∓45	7	60	1	1BB	5" HC
H-15	H∓45	8	60	1	1BB	5" HC
H-2	H∓30	21	100	3	1CL	HC 6"
H-2	H∓30	22	100	4	1CL	HC 6"
H-2	H∓30	26	100	5	4DD	HC 5" or AA
H-2	H∓30	13	100	5	4DD	HC 5" or AA

Time						
From	To	Target	Rounds	FSG	Ships	Remarks
H-2	H±30	14	100	6	2DD	HC 5" or AA
H-2	H±30	15	100	7	2DD	HC 5" or AA

Transfer all FSG on call.

From	To	Target	Rounds	FSG	Ships	Remarks
H-5	H-4	2			LCT(R)	
H-5	H-4	3			LCT(R)	
H-5	H-4	10			LCT(R)	
H-5	H-4	11			LCT(R)	

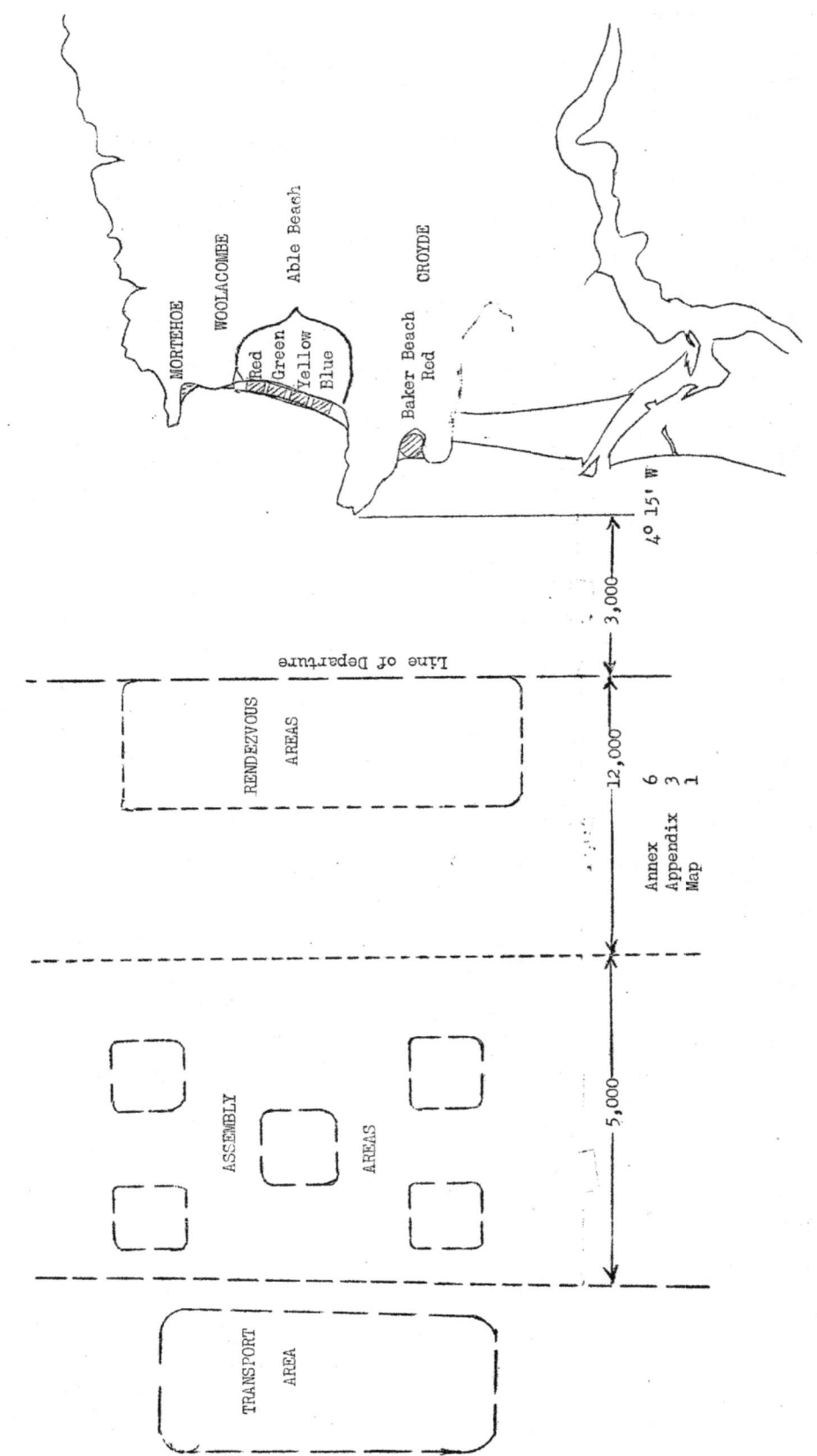

ASSAULT TRAINING CENTER
CONFERENCE

HQ ETOUSA

ANNEX NO. 7 to FO I, SIGNAL

MAPS: Operations Map

1. <u>a.</u> Annex I, Intelligence.

   <u>b.</u> (1) 10th Inf Div (reinforced)

        atchd:
           100th Ranger Regt
           901st Tk Bn
           901st TD Bn
           AA Gp
           501st AA Bn AW
           502nd AA Bn AW
           94th Cml Bn
           Shore Party Gp (See par 2a FO No. 1)

2. The signal communication system during the voyage and the successive phases of the assault will be in accordance with SOP and as further indicated in paragraph 3 below.

3. <u>a.</u> (1) Radio personnel of division, regimental, and infantry battalion echelons will operate sets installed in ships in which their respective headquarters are embarked.

    (2) Wire communication per SOP as early as tactical developments on land permit.

   <u>b.</u> Assault CTs

    (1) Prompt Reports as to "successful landings" will be made to next higher units.

   <u>c.</u> Div Res:
        30th CT

    (1) While in floating reserve will listen in on division command net, 28th CT command net, 29th CT command net (see SOI for call signs and frequencies)

    (2) Special messenger boat service to be established between elements of division reserve.

   <u>d.</u> AA:

    AA Btrys attached to CTs will be assigned one radio set from the Infantry Regiment to which attached to work in CT command ne This is in addition to normal AA Bn, radio nets.

X. (1) Message Centers will detail messengers to remain at the beach to meet and direct members of the next higher unit to their respective CPs.

(2) Watches will be synchronized from official ship time at 2300 hours on D minus 1 day. Thereafter, time signal will be flashed over division command net at 1200 each day.

(3) Radio silence is imposed until H-Hour or it is clear that surprise and security is lost. All nets will be manned on <u>listening watch only</u> from H minus 2 hours. Breaking of silence prior to H-Hour only on direction of CG 10th Division.

4. <u>a</u>. Signal supply dump: to be announced when established.

5. See index No. 1 to SO I.

        By command of Major General _____

                                  X
                                C/S

OFFICIAL:

    Y
    G-3

ASSAULT TRAINING CENTER
CONFERENCE
HQ ETOUSA

Annex #8
Appendix 1

SMOKE PLAN
(AIR)

for Landing Operation, Woolacombe Area.

1. Airplanes (A25) equipped with M-20 smoke tanks, capable of producing a curtain 4000 yards long and with a duration of ten (10) minutes will be used to furnish the initial screen. Three waves of five planes each, with 100% spares to accompany each wave for expected losses, will screen the operation for thirty (30) minutes, from 20 minutes before H-Hour until 10 minutes after H-Hour. Following table indicates the flights and targets of A/C are shown on map over-lay appendix.

AIR SUPPORT FOR SMOKE MISSIONS

Time	No. of A/C	Type of A/C	Load	Objective area	Remarks
H-20	1	A 25	M-20 Smoke Tk	1	Smoke screen
H-20	1			2	" "
H-20	1			3	" "
H-20	1			4	" "
H-20	1			5	" "
H-20	5			Extra	Repl for expected losses
H-10	1			1	Smoke screen
H-10	1			2	" "
H-10	1			3	" "
H-10	1			4	" "
H-10	1			5	" "
H-10	5			Extra	Repl for expected losses

- 1 -

Annex #8
Appendix 2

H Hour	1	1	Smoke screen
H Hour	1	2	" "
H Hour	1	3	" "
H Hour	1	4	" "
H Hour	1	5	" "
H Hour	1	Extra	Repl for expected losses

CML BN

Annex #8
Appendix 2

2. a. 94th Cml Weapons Bn will support the landing operation. Co A is attached to 28th RCT on BAKER BEACH; Cos B and C to 29th RCT, Co B to Inf Bn on ABLE BEACH YELLOW and Co C to Inf Bn on ABLE BEACH GREEN; and Co D to be attached to 30th RCT in floating reserve. The targets and areas assigned to each platoon are indicated on the overlay for Smoke Plan. The Cml Troops, using 4.2" mortars will advance in the initial wave, taking up position about 2000 yards off shore, remaining in that position to support their units with smoke or HE until initial beach objective is taken, except as noted in c. below.

b. The airplanes will furnish the initial screen, when the first wave of the landing craft is about 2½ miles off shore, and maintain the initial screen until ten minutes after the landing. The 4.2" mortars of the Cml Units will be used to put up a smoke screen over any area not covered by the airplanes prior to H plus ten minutes, and after H plus ten minutes the Cml Units will maintain a smoke screen over the entire costal area indicated on the map overlay, until the shore line defenses have been overcome by the attacking troops.

c. The Cml Platoons covering the landings on ABLE BEACH and BAKER BEACH will advance in the first wave and place smoke generators on those beaches for local protection during landings and will also use the 4.2" mortars for firing HE on targets as required during the initial landing operations.

d. After the shore line defenses have been overcome, the Cml Units will be landed and will continue to support their units in the advance with smoke or HE as required, all as indicated in the table below.

ASSIGNMENTS OF 94TH CML BN. IN THE SMOKE PLAN

Annex #8
Appendix #2

Co.	94th Cml Bn Plat.	Sec.	Initial Beach Area Assigned	Unit to Support	Landing on Beach
A	1	1	13	Cos "A" & "B" 28 Inf	Baker
	1	2	14	Cos "C" & "D" 28 Inf	Baker
	2	1	15	Cos "E" & "F" 28 Inf	Baker
	2	2	16	Cos "G" & "H" 28 Inf	Baker
	3	1	17	Cos "I" & "K" 28 Inf	Baker
	3	2	18	Cos "L" & "M" 28 Inf	Baker
B	1	1	7	Co "A" 29 Inf	Able Yellow
	1	2	8	Co "B" 29 Inf	Able Yellow
	2	1	9	Co "C" 29 Inf	Able Yellow
	2	2	10	Co "D" 29 Inf	Able Yellow
	3	1	11	Co "E" 29 Inf	on order CO 2d Bn
	3	2	12	Co "F" 29 Inf	on order CO 2d Bn
C	1	1	1	Co "G" 29 Inf	on order CO 2d Bn
	1	2	2	Co "H" 29 Inf	on order CO 2d Bn
	2	1	3	Co "I" 29 Inf	Able Green
	2	2	4	Co "K" 29 Inf	Able Green
	3	1	5	Co "L" 29 Inf	Able Green
	3	2	6	Co "M" 29 Inf	Able Green
D	1		Floating Reserve	1st Bn 30 Inf	on orders CO 30 RCT
	2			2nd Bn 30 Inf	on orders CO 30 RCT
	3			3rd Bn 30 Inf	on orders CO 30 RCT

ANNEX 8
APPENDEX 3
MAP 1

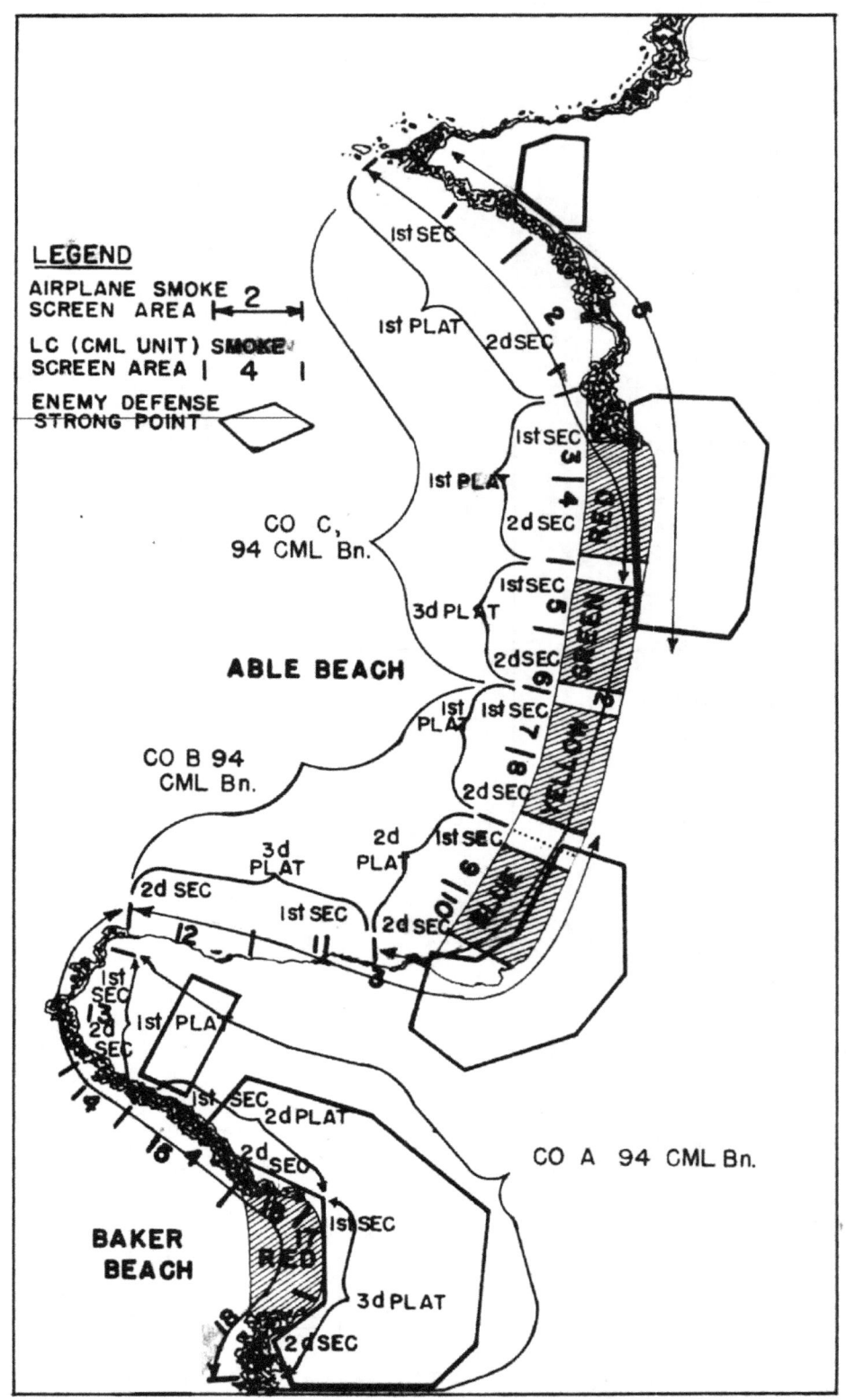

PHASE IV

ADAPTATION OF FM 31-5

HEADQUARTERS
ASSAULT TRAINING CENTER
ETOUSA (PROV)

1 July 1943

FM 31-5. LANDING OPERATIONS ON HOSTILE SHORES

    The Conference on Landing Assault Doctrine has reviewed FM 31-5 for the purpose of adapting it to the specific task of a cross-channel operation against the strongly defended shores of northwest Europe. Such changes considered necessary for adaptation of the FM 31-5 to meet this requirement are indicated herein.

CHAPTER 2

ORGANIZATION FOR LANDING OPERATIONS

SECTION II

BEACH AND SHORE PARTIES

46. BEACH PARTY - omit entire paragraph.

47. SHORE PARTY.-a. The shore party group is the task organization of the landing force for the control of administrative activities at the beach. It is commanded by an army officer, who is on the staff of the senior commander on the beach and is known as the shore party commander.

b. The shore party group consists of a headquarters and a naval beach section and may include any or all of the following sections or such additional sections as may be found necessary: medical, supplies, labor, engineer, military police, chemical, communication, and essential air force units. These sections are organized and trained as a shore party group to secure effective operations at the point of landing during and immediately after the assault. They pass to control of higher commands as soon as the situation warrants.

c. The tasks of the shore party group include:
  (1) Maintenance of liaison between the senior troop commander ashore and shore party group.
  (2) Communication both ship-to-shore and shore-to-shore.
  (3) Maintenance of order
  (4) Control of stragglers
  (5) Direction of traffic and work of prisoners
  (6) Selection and marking of routes inland
  (7) Assignment of operating, bivouac, parking and storage areas for the services using the beach.
  (8) Unloading equipment and supplies from landing craft
  (9) Prompt movement of equipment and supplies from shore
  (10) Decontamination of gassed areas
  (11) Establishment of information and message centers
  (12) Making recommendations as to landing of vehicles and supplies, and the establishment of a supply system
  (13) Evacuation of casualties and prisoners of war to the landing craft.
  (14) Coordinating and effecting movement of air force supplies and personnel to correct location by Air Force Shore Detachments.
  (15) Maintaining warning watch for enemy aircraft by Aircraft Warning Units.

d. The naval beach section is composed of naval personnel who have been assigned and trained with the shore party group. This section is commanded by a naval officer, who is subordinate to the shore party group or shore party commander. The size of the naval beach section depends on the tactical mission assigned to the assault division; it normally consists of boat salvage and repair, engineer, labor and communications personnel.
  (1) The tasks of the naval beach personnel are: reconnaissance of the beach; selection and marking of landing places; marking of hazards to navigation; boat traffic control; aiding retraction of landing craft; emergency boat repairs and salvage; construction of landing facilities and naval communications. In addition to the above, the naval beach section is charged with the prompt despatch of landing craft after unloading.

(C 1, 1 July 1943)

48. LANDING.-a. Depending on the tactical situation and the decision of the army unit commander, the shore party group commander with a small nucleus of the shore parties, lands in one of the leading waves. Shore parties are not landed until required, as the bulk of the working details will not be needed

until large quantities of supplies begin to arrive.

b. The naval beach section personnel may be used to assist boat crews during the movement from ship-to-shore-or-shore-to-shore, likewise the army personnel of the shore party group may be used to assist the boat crews in any way during the movement to the shore.

(C 1, 1 July 1943)

49. INITIAL EVACUATION SERVICE.—a. The shore party is responsible for the evacuation of casualties from the shore to landing craft on the beach. The naval task force is responsible for the evacuation to hospital ships, transports, or the near shore. Medical personnel or ambulance boats are furnished by the navy or as directed in the administrative plan.

b. The beachmaster should have the ambulance boats assembled where they will be less exposed to fire and can be called to the evacuation landing as required.

c. For details of medical services afloat and ashore, see Section VII, Chapter 10.

(C 1, 1 July 1943)

50. PERSONNEL.—a. The shore party group, including naval beach sections, should be organized and made available to the unit organization of the army force which will make the assault landing. Shore party groups are usually attached to a division and may be sub-divided into as many shore parties as required.

b. The number and composition of shore parties is dependent on the size of the landing force and the number of beaches utilized. Labor details in particular will vary with the amount and type of equipment and supplies landed at each beach. In general, one shore party is organized from the shore party group and embarked on the same transport with each assault battalion combat team, or with the landing craft division in a shore to shore movement making up a battalion combat team. If a particular operation does not require all of the shore parties so constituted, the personnel not needed initially may be held in reserve, or used to reinforce the parties on the more important beaches. In the same manner, when beaches are abandoned, shore party personnel thereat are moved promptly to reinforce the parties on beaches to be kept in operation.

(C 1, 1 July 1943)

51. COOPERATION – omit entire paragraph.

(C 1, 1 July 1943)

FM 4

CHAPTER 3

LANDING BOATS

SECTION I

GENERAL

60. SIZE
    xxxx
d. Omit

64. RESERVES. – Boats with large capacity are preferred for transporting reserve battalions which are landed under the protection of other troops. Where it is necessary to land reserve battalions in the second trip of the boats, the movement can be expedited by embarking the reserves on destroyers or other small craft which are moved as close to the beach as safety permits.   (C1, 1 July 1943)

65. – LIGHTERS AND BARGES. – a. Landing of heavy artillery, tanks, vehicles and heavy material is made from standard landing craft with comparatively shallow draft.

  b. Water barges may be required for delivery of water in bulk to the beach during the later phase of landing.

(C 1, 1 July 1943)

SECTION II

STANDARD ARMY AND NAVY BOATS

66. GENERAL. Landing craft of various sizes and characteristics specially designed for assault operations, suitable for landing personnel, supplies, vehicles, artillery and equipment, are available. U.S. Fleet Publication, FTP 207 and 211 give the characteristics of these various craft.

(C 1, 1 July 1943)

CHAPTER 4.

FM 5

SHIP-TO-SHORE MOVEMENT

SECTION I

GENERAL

74. SCOPE.

* * * * * * * *

c. Planning and execution of the movement of those units which operate shore to shore, as distinguished from shore-to-shore are to be additionally considered. The methods and general principles are not to be radically altered.

(C 1, 1 July 1943)

75. SHIP-TO-SHORE MOVEMENT VITAL PART OF ATTACK.

* * * * * * *

(added)
e. The U.S. Fleet Training Publication No. 211 "Ship-to-Shore-Movement" should be consulted as a supplement to this chapter. Familiarity with naval methods and system, including hydrographic and beach markings, is essential for the operation of such a movement.

(C 1, 1 July 1943*

SECTION VII

EXECUTION

112. PREPARATION FOR DEBARKATION. - a. Most landing ships are provided with arrangements such as a gantry for loading the landing craft at the rail and then lowering the craft into the water already loaded. When such mechanical arrangements are not provided, cargo nets are hung over the side of the transport and are used as gangways for debarkation of personnel to the landing craft which are placed alongside the transport. The nets should be sufficiently large to permit four or more men to debark abreast and should reach from the deck to the water line (see fig.11).

* * * * * * * *

(C 1, 1 July 1943)

114. DEBARKATION OF ANIMALS. No application to the movement being considered.

(C 1, 1 July 1943)

CHAPTER 5

OPERATIONS ON SHORE

SECTION II

GENERAL

121. TACTICAL UNITY. - It is imperative that the integrity of tactical units even down to the squad be preserved in the landing. This does not preclude the reorganization of tactical units into assault teams at the time of the designation of an assault force. The new units so formed should be organized prior to or immediately after basic amphibious training, and should maintain their identity throughout the entire specialized assault training for the actual operation, and through the entire assault phase of the operation itself.

* * * *

(C 1, 1 July 1943)

122. NAVAL GUNFIRE, AIR, AND BOAT GUN SUPPORT.

* * * *

d. (added) Consideration should be given to the use of mobile artillery in landing craft to fire on area targets during the approach to the beach.

(C 1, 1 July 1943)

SECTION II
RECONNAISSANCE PATROLS PRIOR TO AND DURING LANDING PHASE

126. GENERAL CONDUCT OF PATROLS
Note change 1, 23 January 1942.

127, 128, 129, 130 and 131.

NOTE.- These paragraphs are useful for information as to landing operations in general, but do not have particular application to the contemplated operation.

(C 1, 1 July, 1943)

SECTION III

BEACHHEAD

139. ESTABLISHING BEACHHEAD. - In a landing operation, troops must clear the beach rapidly. This will apply even though the beach position is heavily fortified or has all types of obstacles.

* * * *

(C 1, 1 July 1943)

140. ADVANCE FROM BEACHHEAD.

* * * *

During this phase, liaison between aviation and ground troops is essential.

(C 1, 1 July, 1943)

FM 7

SECTION IV

SCHEME OF MANEUVER

141. GENERAL.-

* * * * * * *

e. (added) Appropriate force of airborne troops to be determined according to the type objective and opposition to be expected.

(C 1, 1 July 1943)

142. FRONTAGE OF ATTACK. -

* * * * * * * *

b......to take care of this increased front. Sufficient reserves must be kept on hand to insure the exploitation of successes and to continue the attack to the final objective. Airborne troops may be utilized in taking the final objective.

* * * * * * * *

c.
* * * * * * * *

(......   (added). The Air Force participation in the assault contemplates an attack on a relative wide front, and should not be confined to a specific beach or beaches. The Air Force must be operated under centralized control to insure maximum flexibility and proper air attacks.

(C 1, 1 July 1943)

143. INFLUENCE OF LANDING BOATS. - The speed with which troops can be put ashore depends upon the number and type of craft available and the distance to the transport or the near shore (in a shore to shore operation) from the various beaches. The scheme of maneuver therefore must take these factors into consideration. The beach gradient may also affect the utilization of various types of craft.

* * * * * * *

(C 1, 1 July 1943)

SECTION IV

SCHEME OF MANEUVER

144. HOSTILE DISPOSITIONS.

* * * * * * *

(added) Special assault units must be organized and trained to land against strongly fortified beaches.

(C 1, 1 July 1943)

145. LANDING BY ECHELON.

* * * * * * *

In addition, heavy counter-battery fire and combat aviation should be employed to neutralize the enemy batteries.

(C 1, 1 July 1943)

SECTION V

WITHDRAWAL AND REEMBARKATION

* * * * * *

149. WEATHER CONDITIONS

* * * * * * *

Through their meterological service, the Navy and Air Force will furnish the Army with weather predictions. The utmost effort will be made by the Navy to take advantage of favorable weather conditions during the reembarkation.

(C 1, 1 July 1943)

FM 9

CHAPTER 6

NAVAL GUNFIRE

SECTION II

CHARACTERISTICS

158. GENERAL.

* * * * * *

b. * * * *
Naval. It is essential however that naval gunnery officers ashore and afloat, and the landing troops have the same maps, bearing the same military grid and prepared concentrations.
    c. When the naval antiaircraft guns are not required against enemy air operations, their use with a high capacity shell and a time fuse, adjusted for air bursts affords a powerful anti-personnel weapon.
    d. * * * *

(C 1, 1 July 1943)

* * * * * * *

160. MUZZLE VELOCITY AND TRAJECTORY
  * * * * (added). However, the use of accurately adjusted time fire, by virtue of its fragmentation effect, is effective against personnel in defilade, despite the high velocity and flat trajectory.

(C 1, 1 July 1943)

162. TYPES OF PROJECTILES AND FUZES
  * * * * (added). New developments with a high capacity ammunition, a new technique of obtaining air bursts with antiaircraft shells and new methods of fire control, render more effective than in the past for the support of landing operations.

(C1, 1 July 1943)

SECTION III

REQUIREMENTS IN GUNS AND AMMUNITION

170. FIRE SUPPORT GROUPS. Naval gunfire support will be expected to be provided by regular combatant ships (BB, CA, CL, DD), on the basis of several ships in general support and at least one ship in support of each assault battalion.

SECTION IV (C 1, 1 July 1943)

COORDINATION OF FIRE

172. COOPERATION BETWEEN ARMY AND NAVY

* * * * * * *

c. (Added) The effectiveness of naval gunfire depends to a very high degree on the competency of the personnel who control it. To obtain maximum effectiveness in a landing operation, there should be a trained shore fire control party, experienced in operating with troops, and on each firing ship there should be an army artillery officer who has had some training and experience in naval gunfire and who can translate the requirements of the troops into terms of naval gunfire.

(C 1, 1 July 1943)

FM 10

CHAPTER 7

AVIATION

		Paragraphs
Section I.	General.	176-181
II.	Air operations preliminary to landing	182-184
III.	Air operations during debarkation	185-187
IV.	Air operations during advance inland	193-194
V.	Ground requirements to facilitate air operations (added)	194-1/5-194-4/5

SECTION I

GENERAL

176. ADVANCE AIR BASES.-
* * * * * *

b. The seizure of an advance air base for air operations is in some situations the function of the Navy, but in others, it constitutes a separate landing operation for which the necessary landing force is provided by the Army.

c. Airborne and air transported troops may be used for the seizure of an advance air base within operating radius of friendly flying fields. Consideration should be given to dispatch of such troops from carriers when operations from land-based fields are impracticable due to distance. For operations of this character see FM 100-5.

(C 1, 1 July 1943)

177. AIR SUPERIORITY. - a.
* * * * * *

Troop transports and troops in small boats offer concentrated targets for hostile aircraft and are extremely vulnerable to cannon and machine gun strafing, bombing and gas attacks. Even a small opposing air force skillfully handled and not effectively neutralized may disrupt the landing and force a withdrawal. It is therefore essential that the bulk of hostile combat aviation capable of intervening during the landing operations be destroyed or neutralized prior to the approach of the transports and supporting naval units within the transport area. Subsequently, fighter aviation must be prepared to furnish protection against air attacks during the critical landing phase.

b. * * * * * *

(C 1, 1 July 1943)

179. COMPOSITION OF AIR FORCE.-

* * * * * *

In any situation however, the air force should be composed of the classes of aviation which can best accomplish the following missions:
    First, gain and maintain local air superiority.
    Second, preliminary bombardment.
    Third, closely support the landing force and convoy.
    Fourth, furnish necessary reconnaissance and observation, including photographic missions.
    Fifth, lift and support airborne troops.

(C 1, 1 July 1943)

FM 11

SECTION II

AIR OPERATIONS PRELIMINARY TO LANDING

183. PHOTOGRAPHY.-a. Air photographs and mosaics, carefully studied, are of assistance in drawing up final plans for the operation.

b.
* * * * * *

(C 1, 1 July 1943)

SECTION III

AIR OPERATIONS DURING DEBARKATION

* * * * * *

185. PROTECTION OF TRANSPORT AREA. * * * *

b. Submarines and surface vessels are an additional menace during debarkation. Aircraft equipped with bombs and cannon may reduce this hazard. Battleship and cruiser aircraft not required for gunnery observation are also employed to establish an air patrol.

c. **
**
** It is extremely difficult to provide proper air support for a night landing in the presence of an alert hostile air force because the transport area may be illuminated by flares and effectively bombed by the defenders.

(C 1, 1 July 1943)

SECTION IV

AIR OPERATIONS DURING APPROACH TO BEACH

190. GUIDE AIRPLANES.-a. Under exceptional circumstances, and when unopposed by hostile aircraft, guide airplanes may be used.

b.
* * * * * *

(C 1, 1 July 1943)

191. SUPPORT WHEN SHIP GUNFIRE LIFTS.-
* * *
* * *
* * *
* * *
* * The time schedule for its operations is prepared jointly by the air force and landing force commanders and must be coordinated with the naval forces.

(C 1, 1 July 1943)

SECTION V

AIR OPERATIONS DURING ADVANCE INLAND

194. SUPPORT AFTER LANDING OF FIELD ARTILLERY.-

* * * * * *

b. * * * * * *

(5) Transport and other suitable aircraft are utilized to carry airborne or other transported troops to operate against selected objectives.

(C 1, 1 July 1943)

FM 12

SECTION VI (added)

GROUND REQUIREMENTS TO FACILITATE AIR OPERATIONS

194-1/5. GENERAL. Seizure of airdromes at the earliest possible moment is essential. Normally it can be delayed for as long as noon D+1 and then only:
    a. If hostile air threat is negligible, and/or
    b. If land bases within efficient operating range are available to friendly fighters.

194-2/5. CONTROL.-a. A nucleus of an air control center will be established in the assault phase for the purpose of fighter control in order to relieve congestion on the Headquarters and fighter control ships and to provide for the eventuality of these ships being disabled.
    b. The function of advance airfield signal units will include the establishment of airdrome control facilities, homing facilities and a link between advance airfields and Air Force Headquarters afloat and ashore.

194-3/5. DEFENSE.-a. Early in the assault it is essential to bring in certain aircraft warning units. These consist of RDF, ground observers, and other reporting agencies. Early consideration must be given to the landing of ground control interception equipment and personnel. These units will eventually be incorporated into the operational control organization.
    b. Prior planning and early command decisions are required in connection with the ground defense of airdromes:
        (1) In the planning phase of an operation utmost consideration should be given to the employment of air based security battalions for the protection of airfields in order to obviate the necessary diversion of troops from combat units.
        (2) The policy as to command **responsibilities** for defense of an airfield is not yet clearly defined - this responsibility must be definitely prescribed prior to actual operations and is especially important in the case of reinforcements temporarily brought in to assist in defense of an airfield.
        (3) Protection of the seized airfield from air attack **evolves** primarily on antiaircraft artillery. Air based security battalions have no organic antiaircraft artillery, therefore, these weapons must be otherwise provided.
    c. When the beach area is beyond range of light artillery and assumes static condition, the employment of barrage balloons should be considered to assist in its defense.

194-4/5. SERVICE.-a. Aviation engineer battalions, and/or airborne aviation engineer battalion units closely follow the assault troops and are responsible for making captured airfields available for early use.
    b. Advance servicing units are prepared to organize the captured airdrome as a service station and to receive the gasoline, oil and ammunition delivered to the airdrome. They quickly follow the assault forces and while it is not contemplated that they will join in the assault they must be equipped to take care of themselves particularly in the face of a determined counter-attack. These units should be promptly relieved in order that they may initiate the organizing of new airfields.

(C 1, 1 July 1943)

FM 13

CHAPTER 8

SIGNAL COMMUNICATION

		Paragraphs
Section I.	General	195-198
II.	(revised) Communication organization for the landing assault.	200-202
III.	(added) Air support communications during landing assault.	203-207

SECTION I

GENERAL

195. GENERAL

* * * * * * *

b. Each embarked military unit should operate its normal Message Center. Nearly all traffic, both incoming and outgoing, will pass through the ship's communication center, which is operated by Navy personnel. Close liaison must be maintained between these activities. The troop message center will provide the necessary message service between the two agencies.

* * * * * * *

d. The Navy is responsible for ship-to-ship, ship-to-craft, and craft-to-craft communications and generally ship and craft-to-shore. The exception is that radio sets of Command Posts or Rear Echelons still afloat will be manned by military operators.

e. A non-combatant ship must be planned, equipped and furnished with a well trained complement for use as a Headquarters ship.

* * * * * * *

(C 1, 1 July 1943)

196. JOINT PLANS - a. Signal communication plans provide for
(1) The additional personnel such as radio operators and visual signalmen required for transports; control vessels; boat groups, boat divisions, and wave commanders; the beachmaster; fire-control parties and air support parties.

* * * * * * *

(C 1, 1 July 1943)

198. SIGNAL PLANS - Prior to embarkation the following signal plans and orders must be prepared to include the following items which will be in the sequence -
 a. Information:
  (1) Such information of enemy communications and radars as may be essential from the point of view of joint operations.
  (2) Such information of friendly communication and radar facilities as may be necessary for the coordinated action of the joint services in joint operations.
  (3) Such information of the communication and radar systems of the participating services as may be necessary for the understanding by each service of the capabilities and limitations of the communication and radar systems of the other services.
  (4) Such information of the aircraft warning service communication system as is necessary to insure the prompt reception and distribution of the information of approach of hostile and friendly aircraft.

FM 14

(5) Such information regarding air support (Naval and/or Military) as may be necessary to insure that requests for air support will be expeditiously handled.

(6) Such information pertaining to the Command, "Set-up" (that is, location of the headquarters or command) posts of all services as may be necessary for the effective installation and operation of the communication and radar systems.

b. Time:
(1) Designation of the time systems to be used for communication purposes and in heading and text of messages.

c. Precedence (priorities):
(1) Establishment of various degrees of precedence (priority), and appropriate methods of indicating them.

d. Radio:
(1) General radio instructions, as necessary, including radio silence restrictions, etc.
(2) Call signs and frequencies, to include such instructions to cover assignment and distribution of radio call signs and frequencies as may be necessary to:
   (a) Insure coordinated action and avoid interference between the services.
   (b) Provide a guide for all services in distributing the particular items or publications.
   (c) Indicate time when call signs and frequencies become effective.
(3) Instructions for the establishment and operation of such special channels of radio communication as may be required for joint operations, including:
   (a) Assignment of personnel and special equipment when necessary.
   (b) Operating schedules, etc.
(4) Authentication instructions.
(5) Special instructions regarding non-military facilities, including supervision, allocation to various services, and designation of call signs.
(6) Coordinate inter-service requirements.

e. Radar:
(1) Radar search plans, designation of sectors to be searched by individual stations, based on capabilities and limitations of the equipment.
(2) Search doctrine (or policy), standard or special.
(3) Methods of passing Radar information, warning net.
(4) IFF (identification, friend or foe) instructions.
(5) Fighter direction procedures, instructions regarding GCI (ground control interception) and AI (aircraft interception).

f. Radio Intelligence: include such information and instructions as may be necessary for:
(1) Coordinated action of all radio intelligence services.
(2) Distribution (or exchange) of radio intelligence.

g. Visual:
(1) General instructions as necessary, including:
   (a) Restrictions as to the use of visual signal equipment, daylight and darkness.
   (b) Priority of various means (flags, searchlights, etc)
   (c) Meanings of special pyrotechnic signals
(2) Visual call signs:
   (a) Call signs needed for intercommunication between the services.
   (b) Guide for the distribution of call signs to all interested elements.
   (c) Time when call signs are effective
(3) Designation of such visual codes as may be required for joint operations.
(4) Instructions for the establishment and operation of such special visual channels as may be required between elements of the participating services.

h. Wire:
(1) Instructions for installation, operation and maintenance of such wire communications as may be needed, including any restrictions.
(2) Organization of nets.
(3) Call signs needed for intercommunication between participating services.
(4) Special instructions regarding non-military facilities, including supervision, allocation to various services, designation of call signs.

i. Messenger Service: instructions for the operation of a messenger service between various elements which should state schedules, means of transportation, and whether officer or enlisted man.

j. Recognition signals:
(1) Recognition doctrine (or policy)
(2) List of the types to be used, day and night, with statements of time when each becomes effective.
(3) Recognition signals (surface craft, submarines, aircraft and ground forces:
  (a) Type or types to be used.
  (b) Prescribed signals and maneuvers, including those used by aircraft approaching surface craft.
  (c) Identifying marks on ships, special flags lights, etc.
(4) Guide for distribution.

k. Communication security and cryptanalytic activities: instructions to insure the coordination of the communication security activities, and the prompt exchange of technical information between cryptanalytic elements.

l. Codes and ciphers:
(1) Systems to be used, effective dates and restrictions
(2) Guide for the distribution of codes and ciphers.

m. Equipment and personnel:
(1) Instructions concerning liaison between services and the training and assignment of special personnel.
(2) Considerations of equipment of the various services, the limitations of use in areas of operation, available replacements, interchangeability and spares.

n. Grid system prescribed.
(added)   (C 1, 1 July 1943)

199. CONSIDERATIONS EFFECTING COMMUNICATION MEANS. - a. Amphibious operations present communications problems which differ in many respects from those encountered in other types of fleet or field operations. These different problems impose a heavy burden on the normal communication agencies. Special training is required and additional personnel and special equipment must be provided.

b. The nature of amphibious operations is such that great dependence must be placed on the use of radio to provide adequate and efficient communication facilities which are prerequisites to success.

c. With radio assuming the largest share of the communication burden, the limitations placed on the use of wire must be recognized.
(1) Wire is completely impossible in the early stages of the assault, and its later use must depend entirely on the development of the situation.
(2) The timely use of wire communication facilities depends upon the keen observation and good judgment on the part of the Signal and Communication Officers concerned. When the situation has developed to where wire can safely and profitably be employed, this means should be effected to relieve the radio channels and thus provide additional operational security.

(C 1, 1 July 1943)

FM 16

SECTION II (revised)

COMMUNICATION ORGANIZATION FOR THE LANDING ASSAULT

200. GENERAL - A standard system of radio nets for the progressive phases of the debarkation and assault operations are contained in four phases as outlined herein.

(C 1, 1 July 1943)

201. RADIO NETS FOR THE PROGRESSIVE PHASES OF OPERATIONS - a. Phase 1. - (1) This phase covers the following evolutions: debarkation of troops, loading of boats, organization of the boat waves, movement of waves to rendezvous area, movement of combat group to line of departure, and final movement to the beach.

(2) During this phase radio silence is normal and communications are restricted to visual means (arm and flag signals during daylight, and carefully screened lights in darkness). However, military tactical radio nets are manned in complete readiness for instant use, should the element of surprise be lost or an emergency arise. Continuous listening watches are maintained at all times.

(3) From the standpoint of efficiency of radio communication it is highly desirable that certain commanders and special communication teams be embarked together in certain boats as follows, in order to establish a standard system of radio nets.

(a) The combat Group Commander with the Landing Team Commander.
(b) The Assistant Combat Group Commander with the Landing Team Executive officer.
(c) The shore fire control party boat adjacent to that of the landing team commander in order to be able to call for direct fire while the landing force is waterborne.
(d) The air liaison party in a boat adjacent to that of the landing team commander if direct air support is provided.
(e) The shore party commander and the beachmaster should be embarked in adjacent boats of the same wave.

(4) Special radio nets: (a) When conditions require the provision of a combat Flotilla Commander a special net is provided which will include the flotilla commander, each of his combat group commanders and transport division commanders. This net is omitted when flotilla organization is not used. (See diagram 5).

(b) The control vessel must be prepared to operate in the combat group commander's circuit, the commander's control vessels circuit and the fire support group supporting the beach. The commander of the transport division in the headquarters ship will listen in on the control vessel circuit. The Division Headquarters ship should operate a transmitter on this circuit if available, and the naval force commander will maintain the listening watch on this circuit during the initial stages of the ship-to-shore movement. If a guide plane be employed the boat group commander and the combat team commanders will maintain listening watches on the guide plane frequency. The boat group commander will also operate a set in the combat team net. The frequency for this net must be prescribed by a higher echelon.

(5) Accompanying diagrams for the radio nets and circuits used in Phase 1 are listed below:

Boat Control Circuit	Diagram 1
Transport Division Command Circuit	Diagram 2
Direct Support Aircraft Circuit	Diagram 3
Combat Team Net	Diagram 4
Boat Group Control Circuit	Diagram 5

FM 17

b. Phase 2. The second phase of the landing attack is the period when landing teams are ashore and combat team communications are still afloat. During this phase the radio, messenger, visual and sound systems are normal. Accompanying diagram 6 is a consolidation of the various circuits and nets during this phase.

c. Phase 3. In this phase of the attack, the several combat team HQs have been established ashore. The initial landing teams have proceeded further inland. All circuits remain unchanged except that some circuits now terminate on means which formerly were afloat.

d. Phase 4. In the fourth and final phase of the landing attack the commander of the landing force has established headquarters ashore. The shore party is in visual communication with the unloading transports and cargo ships. Naval, air and main fire support continue as long as they are needed. In this phase all signal communications will approach Intra Service normal systems. Wire, radio, messenger, visual and sound will have been installed and in operation and modifications will be initiated in anticipation of the consolidation. The wire system will be initiated during this phase and developed as early as the situation permits in order to tie the Division Hq's into the multi-channel radio system. The accompanying diagram 7 is a consolidation of the various circuits and nets used during this phase, prior to installation of the wire system.

(C 1, 1 July 1943)

202. PERSONNEL AND EQUIPMENT. - a. Signal communication plans provide signal personnel and equipment to replace losses during the ship-to-shore movement as well as during the early subsequent operations ashore, and to meet the needs of the situation for special eqipment. This applies particularly to the assault combat teams and the shore parties.

b. All signal communication equipment of the assault combat teams and shore parties is portable. Adequate equipment for laying wire is provided. To guard against total loss of equipment through the sinking of a boat, it is desirable to provide duplicate sets of equipment in separate boats. The lack of transportation and the disorganization likely to prevail during the early operations ashore impose unusual demands on the signal communication personnel of the assault combat teams. This personnel is frequently augmented by the attachment of regimental or higher unit signal communication personnel.

c. In every case, wire to replace that expended by assault combat teams and reel units, with transportation therefor, for the mechanical laying of wire are sent ashore and pushed forward at the earliest opportunity.

(C 1, 1 July 1943)

PHASE I   BOAT CONTROL CIRCUITS

RADIO -- SCR-511 or TBY

NAVY CIRCUIT

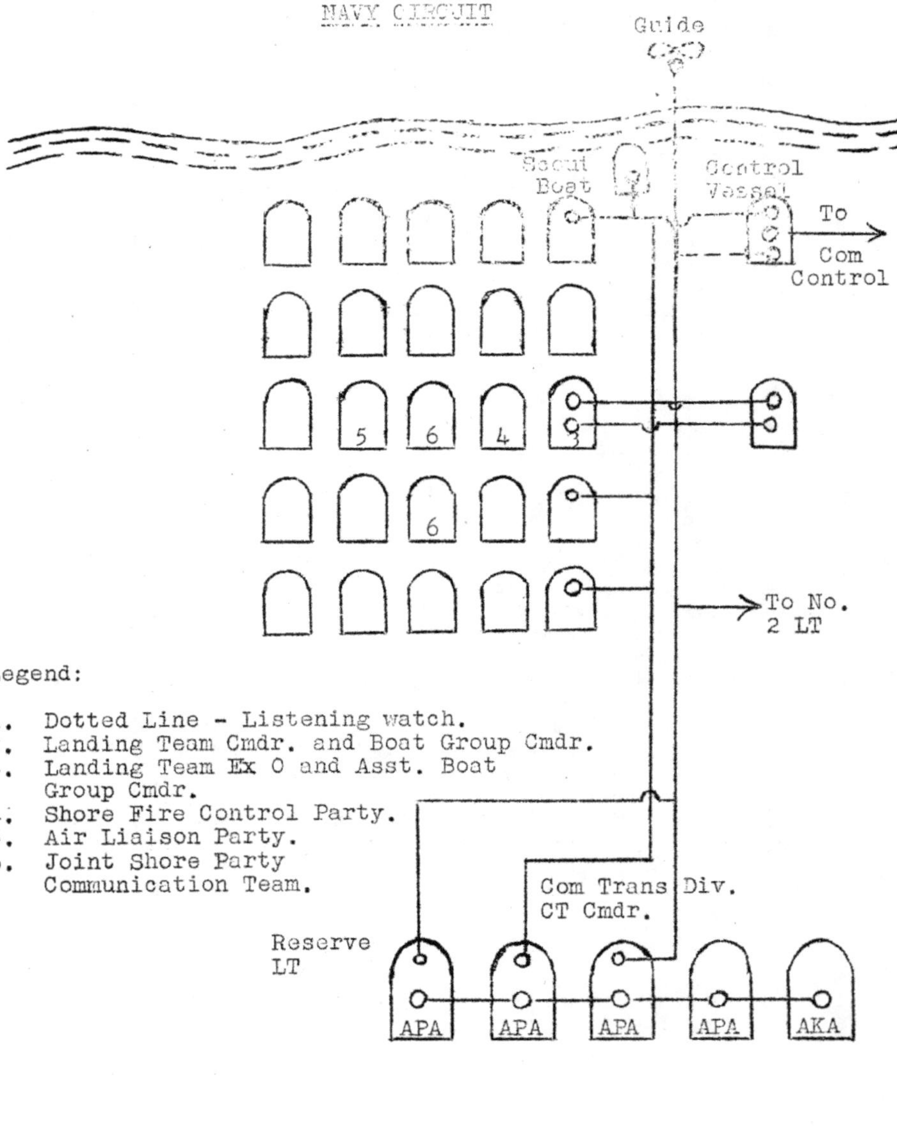

Legend:

1. Dotted Line - Listening watch.
2. Landing Team Cmdr. and Boat Group Cmdr.
3. Landing Team Ex O and Asst. Boat Group Cmdr.
4. Shore Fire Control Party.
5. Air Liaison Party.
6. Joint Shore Party Communication Team.

DIAGRAM #1

PHASES I - IV

TRANSPORT DIVISION COMMAND CIRCUIT
RADIO - Ship's Radio
NAVY CIRCUIT

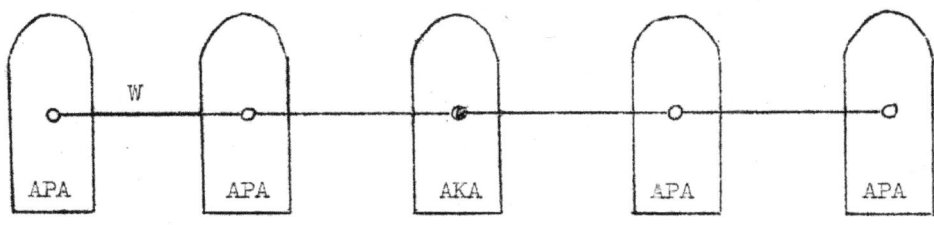

Comtransdiv
CT Comdr.

THIS IS AN EMERGENCY CIRCUIT

This circuit gives the Transport Division Commander control of the ships under his command. In practice two paralleled circuits may be employed, one a voice circuit for emergency use, the second a key circuit for other purposes. Visual signalling is used whenever possible.

DIAGRAM #2

PHASE I

DIRECT SUPPORT AIRCRAFT CIRCUIT

RADIO - SCR-193

JOINT CIRCUIT

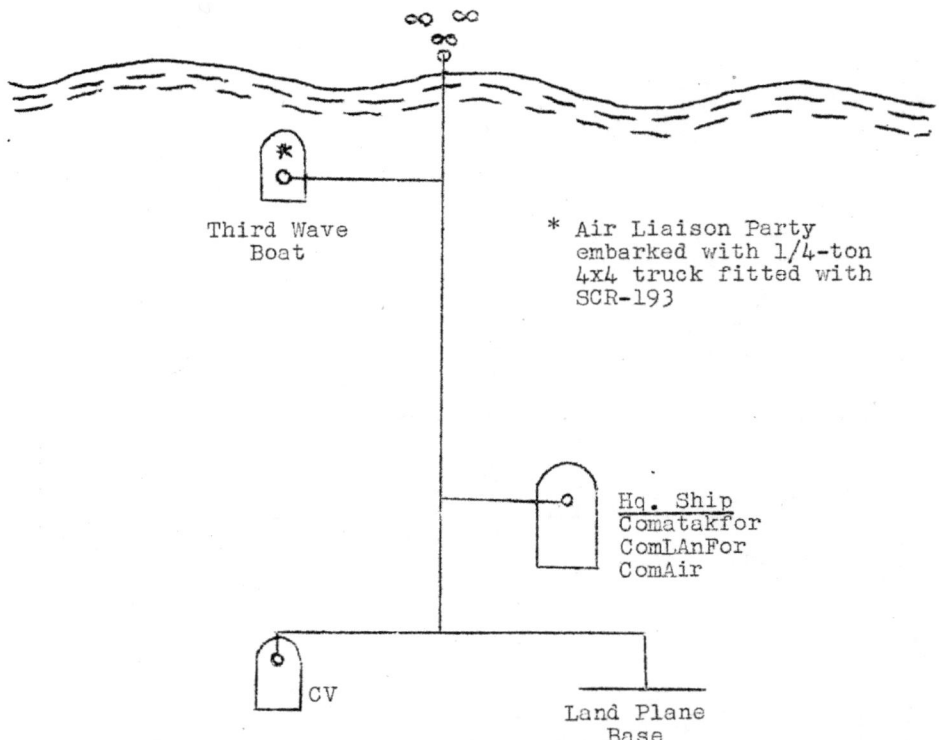

* Air Liaison Party embarked with 1/4-ton 4x4 truck fitted with SCR-193

The purpose of this circuit is to afford communication between the supporting aircraft and the assault troops. The Air Liaison party may go ashore with a Landing Team; or later with the CT Headquarters.

DIAGRAM #3

PHASE I

COMBAT TEAM NET
RADIO -- TBX or SCR - 284
ARMY OR MARINE NET

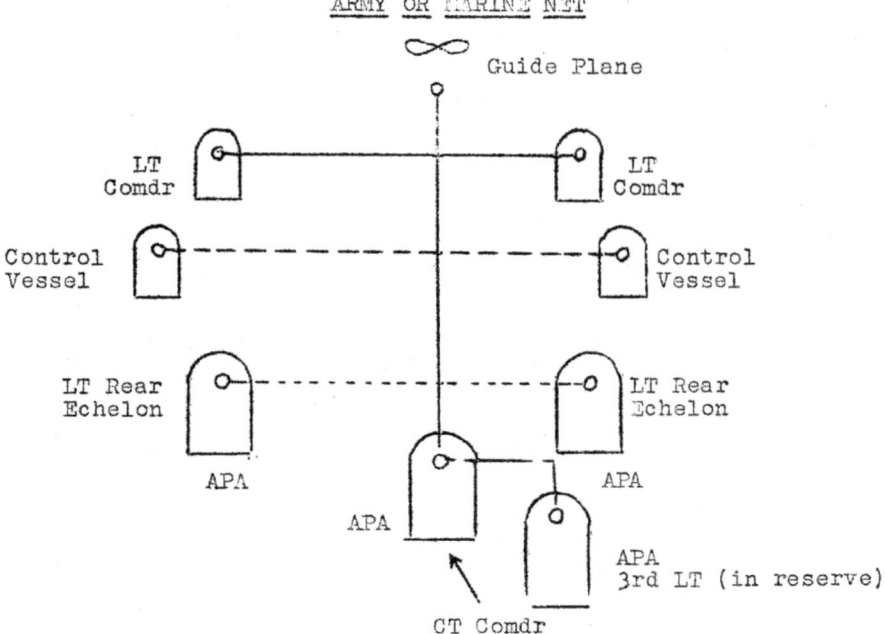

Dotted lines indicate listening watches only.

    Each Transport carries a battalion Landing Team. This circuit (net) is between the Combat Team commander, on the Trans Div Flagship and the Landing Team commanders for tactical control. One Landing team is normally kept in reserve either in the transport or in boats. Each Landing Team communication platoon is divided into two echelons, a forward and a rear, the latter maintaining a listening watch on this net. The several landing teams operate on different beaches. A guide plane may be on this circuit in order to assist the boat waves to the beaches. Control vessels should listen for information.

DIAGRAM #4

PHASE I & II   BOAT GROUP CONTROL CIRCUIT

RADIO - TBY or SCR-511

NAVY CIRCUIT

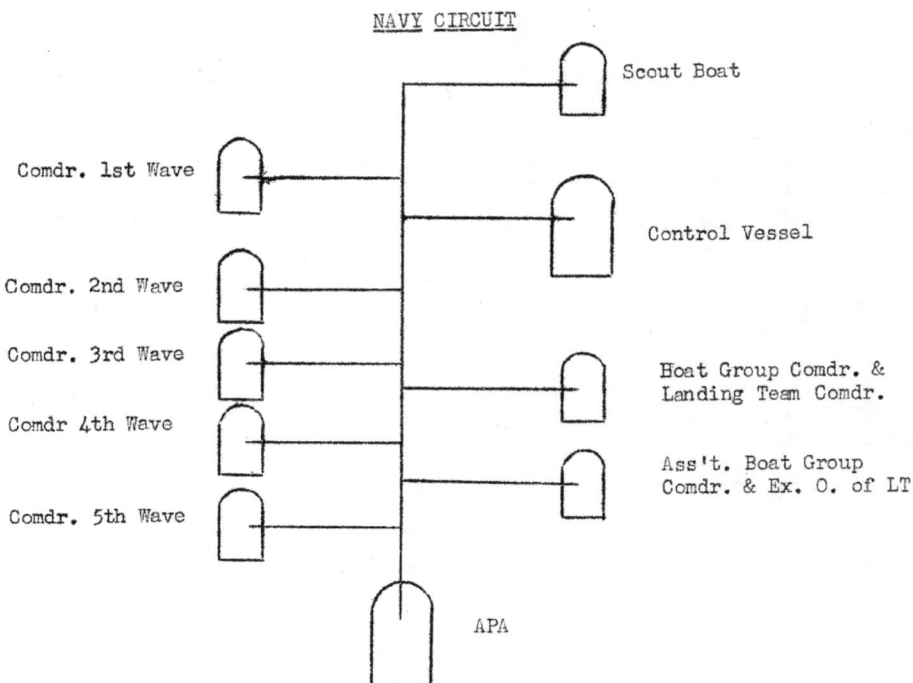

Note: This diagram doesnot indicate the correct position of
the boats relative to the waves, but merely indicates
the communication net.

DIAGRAM # 5

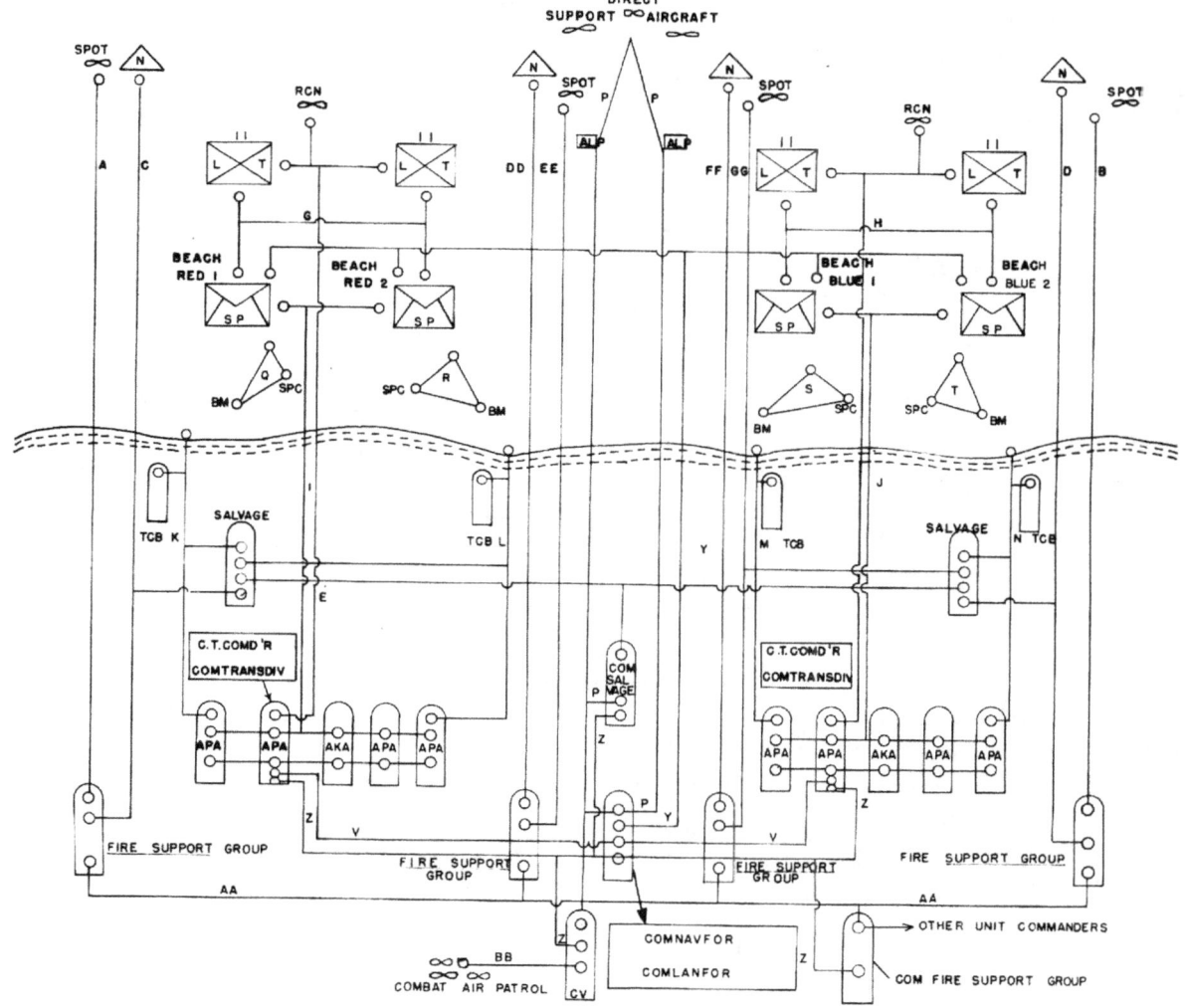

PHASE IV

AMPHIBIOUS RADIO CHANNELS

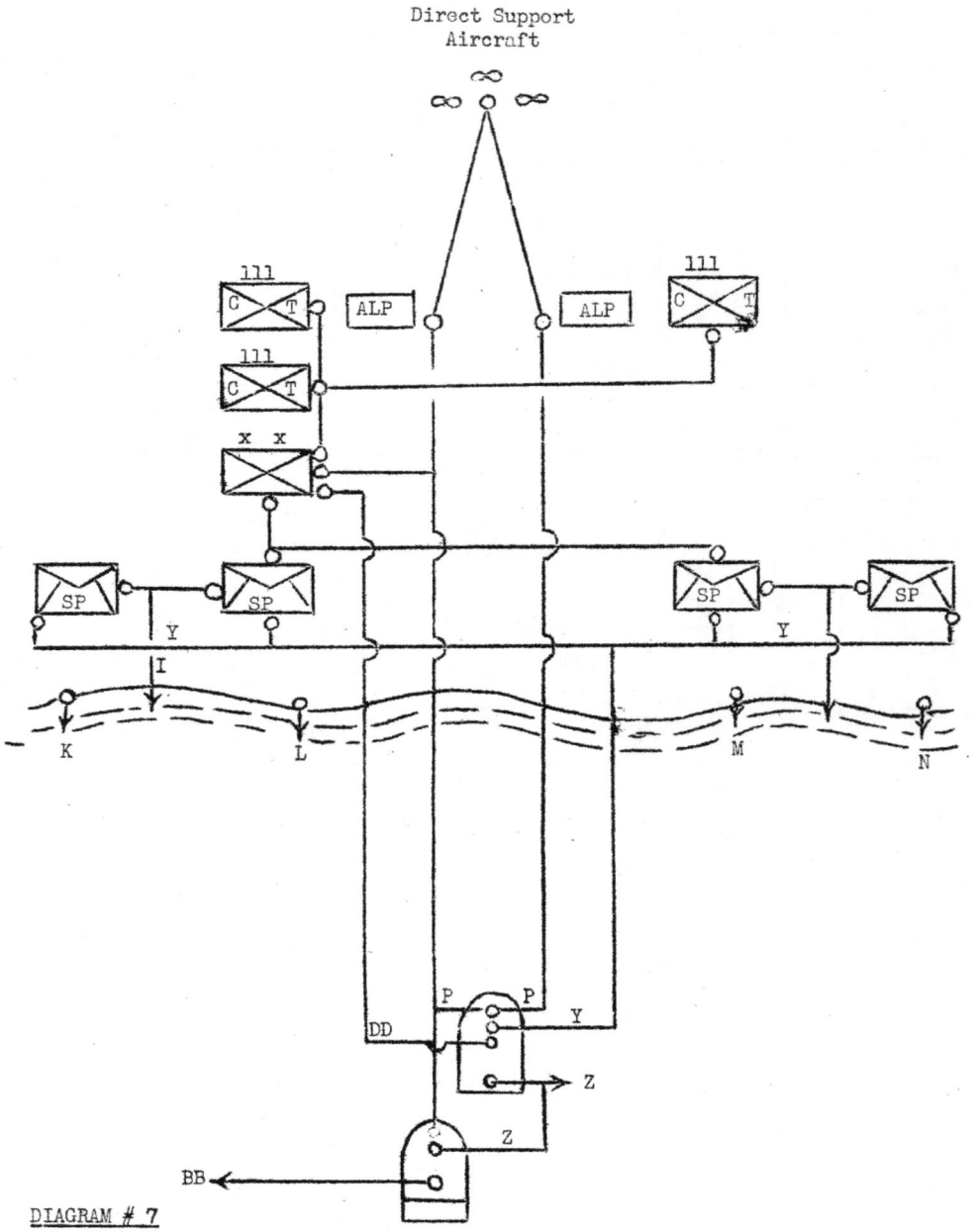

DIAGRAM # 7

SECTION III (added)

AIR SUPPORT COMMUNICATIONS DURING LANDING ASSAULT

203. GENERAL. The principles governing communications for air support are of similar nature to those encountered in the communications phases of ground operations, the noticeable difference being brought about by the strategic employment of air formations in support of the particular operation.

204. CONTROL COMMUNICATIONS. In all air amphibious operations, the Air Force Commander exercising control from Air Force Headquarters on the near shore, requires certain communication channels (see diagram 8) to the Headquarters ship for the control and direction of aircraft, and to provide liaison between the assault force Headquarters Ship and the Air Force Headquarters ashore. In addition it will be necessary to provide facilities for the reception of Radar information broadcast from the friendly shore and later from the far shore.

205. FIGHTER CONTROL SHIP. If the operation permits, consideration should be given to the employment of a Fighter Control Ship to relieve congestion aboard the assault force Headquarters Ship, aircraft and ground units ashore. This facility is desirable, if practicable, to provide duplicate communication facilities and to relieve congestion and control activities aboard the assault force Headquarters Ship (see diagram 9 and 12).

206. COMMUNICATIONS ASHORE. The air force portion of the assault force headquarters ashore must be provided with communication facilities to the assault force headquarters ship and aircraft and ground units ashore. The assault force headquarters ship and/or fighter control ships should be used until the build-up ashore provides adequate communication facilities to sustain the communications required. The build-up of the Radar network ashore should begin by the landing and siting of light warning sets at the earliest opportunity in order to give maximum cover to the landing areas. Ground control interceptor equipment must be landed as soon thereafter as possible to establish nucleus fighter control. Control of low cover fighters by day to be retained by the assault force headquarters ship and/or fighter control ships until the build-up of communication facilities ashore permits the transfer of this control thereto. Control of high altitude fighters to be retained by the more permanent system on the friendly shore until adequate sector control organizations exist on the beachhead (see diagram 10).

207. COMMUNICATIONS ASHORE FOR ADVANCE LANDING GROUNDS. When advanced landing grounds are established on the beachhead they will require communications to aircraft and to the assault force headquarters. Air force elements of the assault force headquarters ashore to be equipped with communication facilities to enable them to control aircraft based on such advanced landing grounds. Additional Radar equipment and ground observers to be integrated into the overall scheme at this point, with a view to their incorporation into the sector fighter control organization. (see diagram 11).

(C 1, 1 July 1943)

EXTRA COMMUNICATIONS REQUIRED AT SHORE

BASES IN SHORT RANGE OPERATIONS

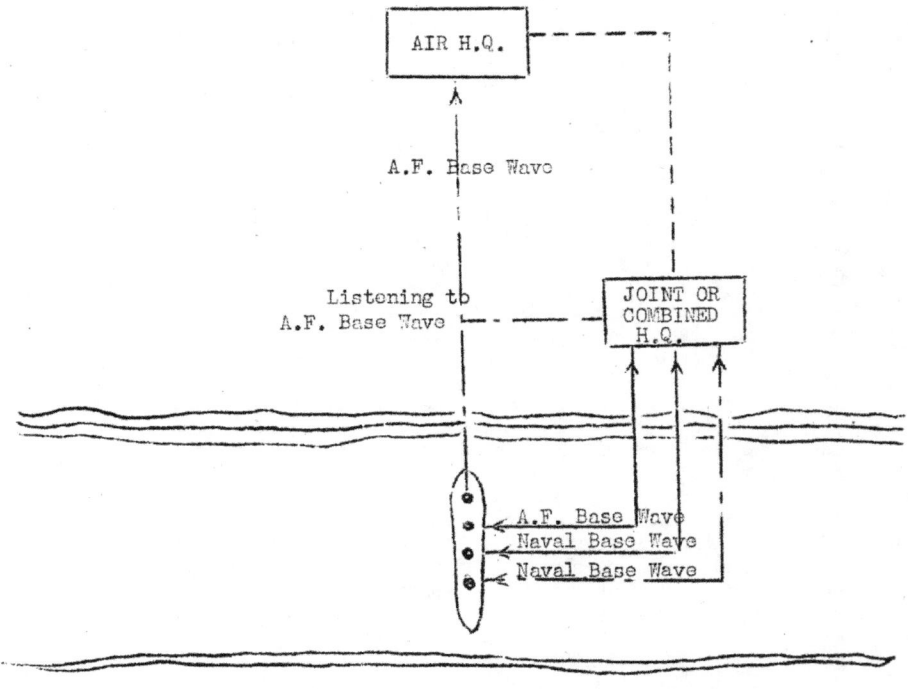

Note
Naval channels shown may by arrangement
act as alternate routes for Air Forces
Traffic

LEGEND
———————— R'T
— — — — W'T
- - - - - - L'L

CHANNELS	APPROX RANGE REQ
AF BASE WAVE H'FW'T	UP TO 100 MILES
AF BASE WAVE VHER'T	UP TO 100 MILES

DIAGRAM # 8

COMMUNICATIONS IN H.Q. SHIP FOR SHORT RANGE
OPERATIONS (DESTROYER OR SIMILAR VESSEL)

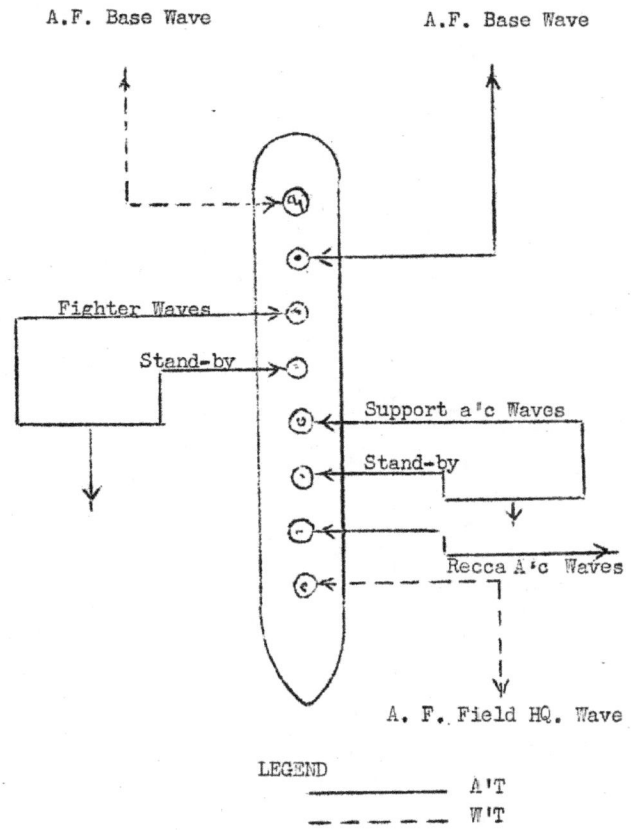

CHANNEL	APPROX RANGE REQ.
AF BASE WAVE H'FW'T	UP TO 100 MILES
AF BASE WAVE VHF W'T	UP TO 70 MILES
FIGHTER WAVES	STRONG SIGNAL
FIGHTER WAVES STAND BY	
SUPPORT A'C WAVES	
SUPPORT A'C WAVES STAND BY)	
RECCE A'C WAVE	UP TO 70 MILES
ASSAULT FORCE HQ WAVE	UP TO 30 MILES

DIAGRAM # 9

SKELETON AIR FORCE COMMUNICATIONS
REQUIRED BY AN A.F. FIELD H.Q.

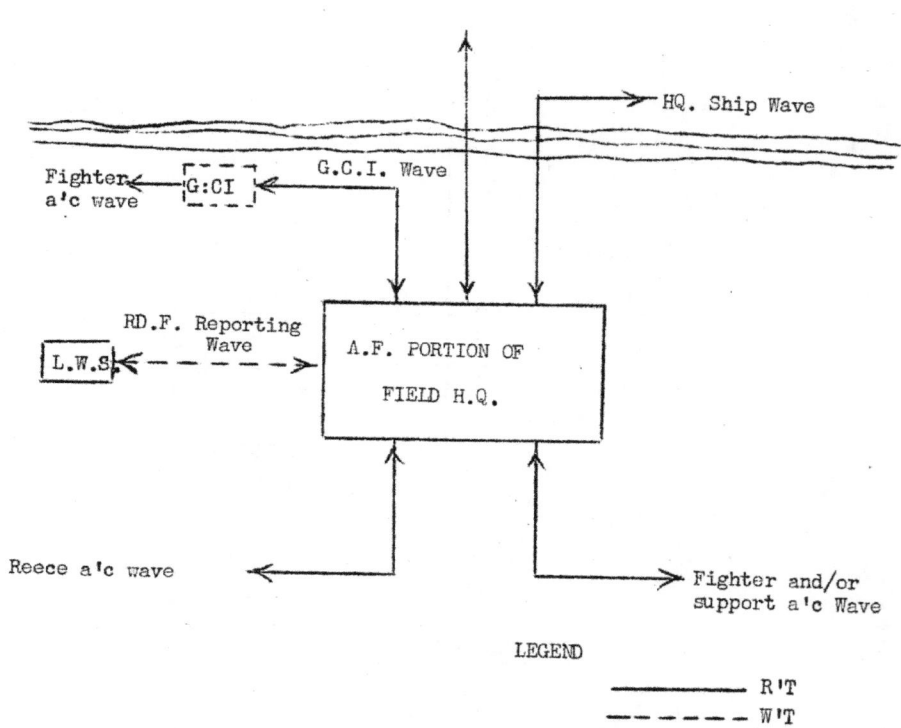

DIAGRAM # 10

COMMUNICATIONS FOR AIR FORCE FIELD H.Q.
ASHORE - INITIAL A.L.G. COMMUNICATIONS
BEFORE ESTABLISHMENT OF SOR'FR ----

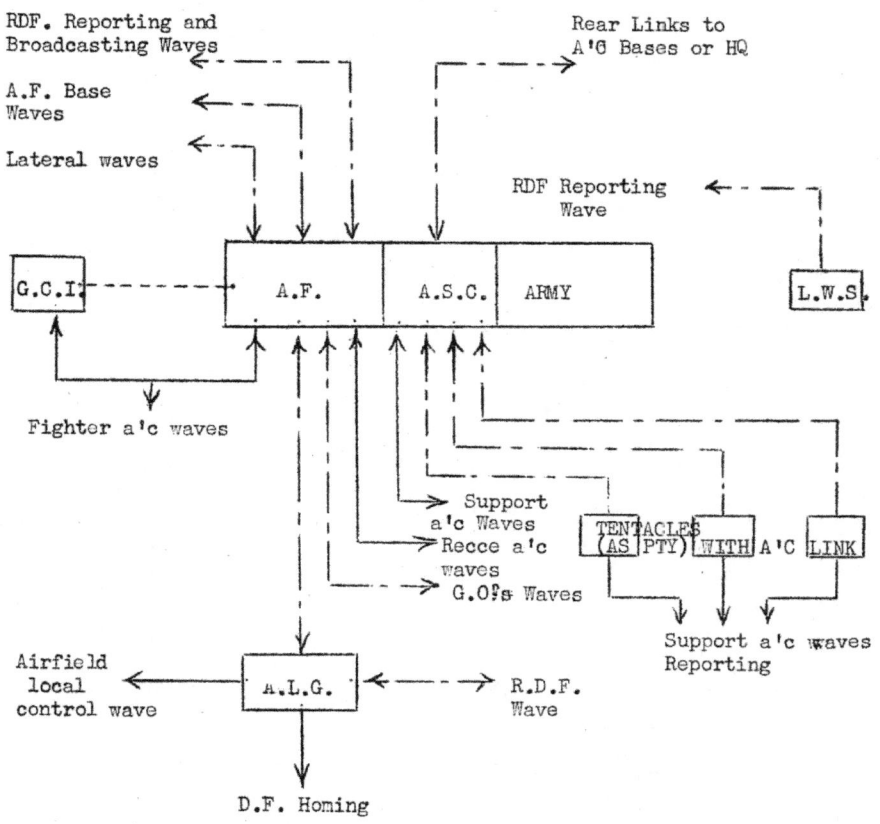

CHANNEL	Approx Range Required
Lateral Waves	Up to 40 Miles
A.F. Base waves	Up to 500 Miles
R.D.F. Reporting	---
Broadcasting wave	---
Fighter a'c waves	---
Rear links to a'c bases or HQ	Up to 100 miles
Air support waves	Up to 50 miles
Support a'c waves	---
Recce a'c waves	---
ALG Waves	Up to 30 miles
Ground Observers Wave	Up to 50 Miles

DIAGRAM # 11

COMMUNICATIONS FIGHTER CONTROL SHIP

USED IN CONJUNCTION WITH HQ SHIP

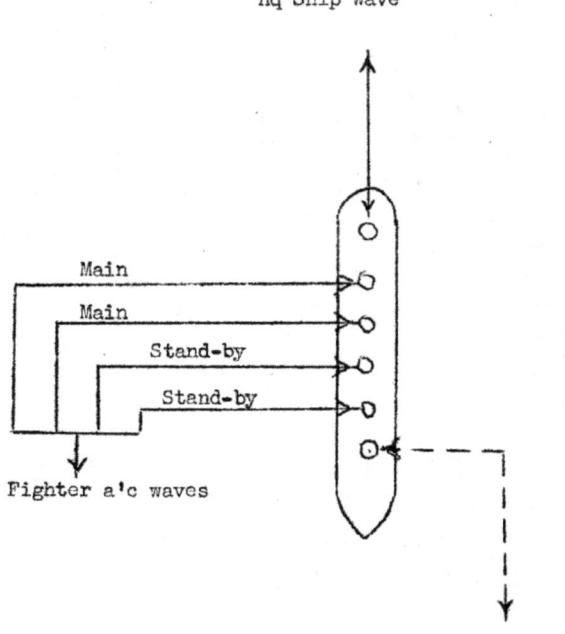

LEGEND

——————— R'T
— — — — — W'T

CHANNEL	APPROX RANGE REQ.
Inter-ship wave	Up to 30 miles
Fighter A'C Waves	--
A.L.G Wave	Up to 50 Miles

DIAGRAM 12

FM 31

CHAPTER 9

FIELD ARTILLERY, ANTIAIRCRAFT PROTECTION, TANK AND ENGINEER UNITS, AND CHEMICALS.

SECTION I

FIELD ARTILLERY

208. EMPLOYMENT OF FIELD ARTILLERY

* * * * * *

b. * * * * * *

(added) The Air Force will also provide a considerable measure of support in accordance with a prearranged plan.

c. Normally in a landing operation field artillery must reach the beach before it can go into action. This factor together with the necessity of reinforcing or relieving naval guns at the earliest possible time, makes it necessary to employ field artillery with great boldness. Divisional light artillery can add considerably to the volume of supporting fires while still afloat, which is possible when fired from special craft. Every advantage must be taken to get the artillery ashore at the earliest moment and in position to assume its primary role of close support of the infantry. Substituting self-propelled guns, both 105mm and 155mm, for organic towed guns renders definite advantages to the assault division.

d. * * * * * *

(C 1, 1 July 1943)

SECTION III

TANK UNITS

216. EMPLOYMENT.—a. Tank units are particularly valuable in a landing operation. They are effective against beach defense machine guns and barbed wire. They assist the advance of assault units during the period when field artillery support is lacking, and are used in later phases in attacks against especially stubborn resistance and counterattacks. In the assault of a strongly fortified coast, tanks are considered as heavily armored assault guns necessary in the reduction of the fortifications. Due to mines on the beaches and extremely heavy defensive fire, the tanks in the leading waves will probably not be able to leave the beaches, but in addition to the fire power they give before and even after losing their mobility, they will also be of considerable value as a shield for the infantry in the leading waves.

b. * * * * * *

(C 1, 1 July 1943)

FM 32

CHAPTER II

SPECIAL TRAINING FOR LANDING OPERATIONS

SECTION III

JOINT SIGNAL COMMUNICATION TRAINING

284. TRAINING OF PERSONNEL. a. All communications personnel should be trained in the signal communication means to be used with the ground units to be supported during the landing. Such training should be in accordance with FM 24-6 and FM 24-10.

b. This training should be effected by formation of the shore party communication teams of military and navy personnel to receive joint training in advance of the contemplated operation. Such joint training should be literally in the same classroom, at the same time, and from the same instructor.

c. Shore fire control parties and air support parties must likewise receive joint training in their respective specialities.

(C 1, 1 July 1943)

286. NAVAL LIAISON AND COMMUNICATION DETACHMENTS.

* * * * * * * * * * *

Note: Use of pyrotechnics should be reserved as emergency means and not relied upon to convey important information, except as a last resort.

(C 1, 1 July 1943)

FM 33

APPENDIX II

TYPES OF NAVY SHIPS AND AIRCRAFT

1. SHIPS

* * * * * * * * * * * *

Auxiliary

* * * * * * * * * * *

       Attack Cargo Ship (Combat loaded) - ARA
       Attack Transport Ship (Combat loaded) - APA
       Headquarters Ship   - AGC

(C 1, 1 July 1943)

APPENDIX III

SMALL BOAT TYPES

* * * **** * * * * * * * * *

NOTE: Boats types included herein are of no particular interest in this operation. Information on types to be used is available in U.S. Fleet Training Publication No. 207.

(C 1, 1 July 1943)

PROPOSED TRAINING CIRCULAR

ASSAULT TRAINING CENTER
CONFERENCE

HQ ETOUSA

1 July 1943

PROPOSED TRAINING CIRCULAR

A rough draft of a proposed Training Circular on the "Attack of a Fortified Beach", prepared by various committees of the Conference, is presented herewith. The principles set forth are considered to be fundamental, and form a basis for further revisions now being prepared by the Assault Training Center Staff for official consideration.

As new techniques and equipment of proven worth are made available, they will be recommended for inclusion.

ASSAULT TRAINING CENTER

CONFERENCE

HQ. ETOUSA.

DRAFT OF A PROPOSED TRAINING CIRCULAR.

SUBJECT: ATTACK OF A FORTIFIED BEACH.

SECTION.

I. General
    1.) Scope
    2.) Definitions

II. Organization of a Beach Defense Area

III. Unusual Characteristics

IV. Weapons and Means Available

V. Organization

VI. Execution of the Attack
    1.) Sequence of Operations
    2.) Assault Principles
    3.) Reduction of Fortifications
    4.) Example
    5.) Time of Attack

VII. Special Training

VIII. Appendix

I. GENERAL

1.) <u>Scope</u>: This training circular embodies a consideration and tratment of as much of the landing phase of "Landing Operations on Hostile Shores" (FM 31-5) as pertains to the seizure of strongly organized beach defenses. Although section 144 of that manual states that, "Beaches strongly organized for defense are avoided, if possible, in the initial landing", circumstances may nevertheless require such an assault. This training circular describes the characteristics of strongly fortified beach areas of German design, considers them in the light of existing doctrine described in "Attack on Fortified Positions", TC No. 33, Mar/43, and sets forth techniques to overcome installations to be encountered in such defenses. It further illustrates the application of these techniques in a tactical situation.

2.) <u>Definitions</u>: THE SEIZURE OF AN ORGANIZED BEACH DEFENSE, for the purpose of this circular, is construed to mean the successful landing of assault elements within appropriate beach areas, together with the reduction of organized positions defending these beach areas.

A STRONGLY FORTIFIED BEACH AREA is a defensive area organized for all-around defense. It will usually consist of minefields, barbed wire, seawalls, the other obstacles located on or close to the beach, covered by direct fire from machine guns, antitank guns and rifles, located in concrete pillboxes or open emplacements, and by supporting fire from artillery and mortars located within 4000 to 5000 yards from the beach.

II. ORGANIZATION OF A FORTIFIED BEACH AREA.

1.) <u>German Policy</u>: a.) The German policy for the defense of France and the Low Countries is to hamper an invading force before it reaches the coast, and, should it reach the coast, to prevent the invaders from crossing the beaches. Their coastal artillery is sited well forward and will engage ships and craft before they can land. Should craft reach the beaches, the German aim is to hold the invaders on the beaches by obstacles, while they are destroyed by fire from well protected localities enfilading these obstacles. Local reserves are used to reinforce the fixed beach defense positions. They are not used to man defense positions further inland. There are main mobile reserves for counter attack. Thus, should an invading force be successful at the beaches (the main line of resistance), the German plan is to stage an armored counter-attack before this force can consolidate a beachhead and get enough troops ashore to offer adequate resistance.

b.) A feature of the German coastal defense system is the large employment of artillery and automatic weapons of all kinds. In addition to heavy and light coastal defense guns and mobile railway guns which are mainly in the vicinity of the ports, there are:

(1) Land batteries of medium caliber, fitted with instruments to engage ships at sea and, in some cases,

also able to fire on the beaches;

(2) Antiaircraft and antitank guns sited where they can, if necessary, engage craft attempting to land;

(3) Field and medium artillery, sited to bring fire to bear on beaches where a successful landing has been made, but from such a distance that the guns themselves will not become involved in the fighting for the beaches. There are on the average some 200 guns of this nature employed to cover a divisional sector.

c.) The German Air Force will support the defense at all stages. A system of airfields has been developed that will allow air support to be concentrated at any point on the coast.

d.) The Germans attach great importance to the defense of ports. They believe that an invading force cannot maintain itself over beaches, and the capture of a port thus is vital. There has been prepared on the landward side of all important ports a defensive perimeter consisting of antitank ditches (sometimes combined with flooding), continuous wire belts, minefields, road blocks and infantry strong points, with trenches, pillboxes, shelters and MG and field gun positions. Quays, jetties, moles and cranes are prepared for demolition.

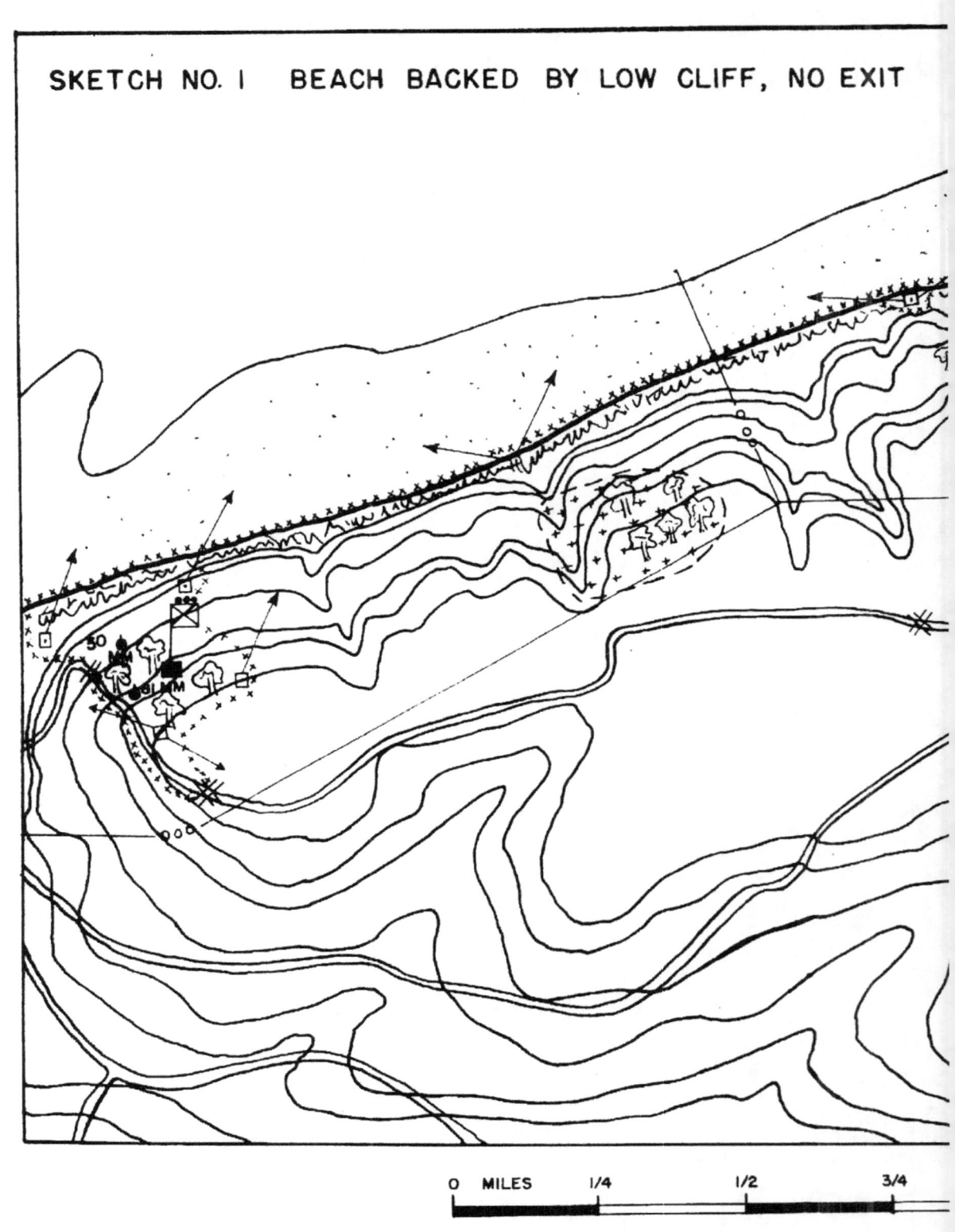

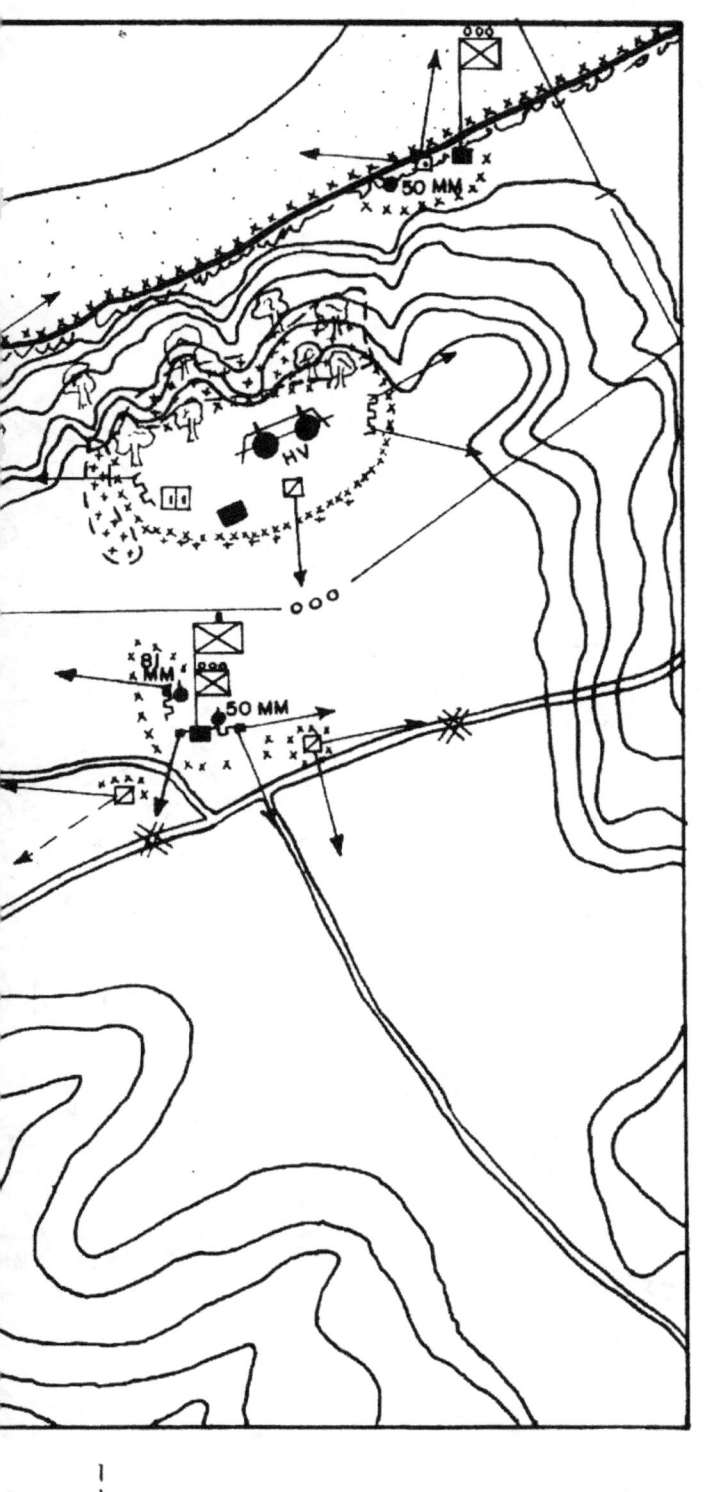

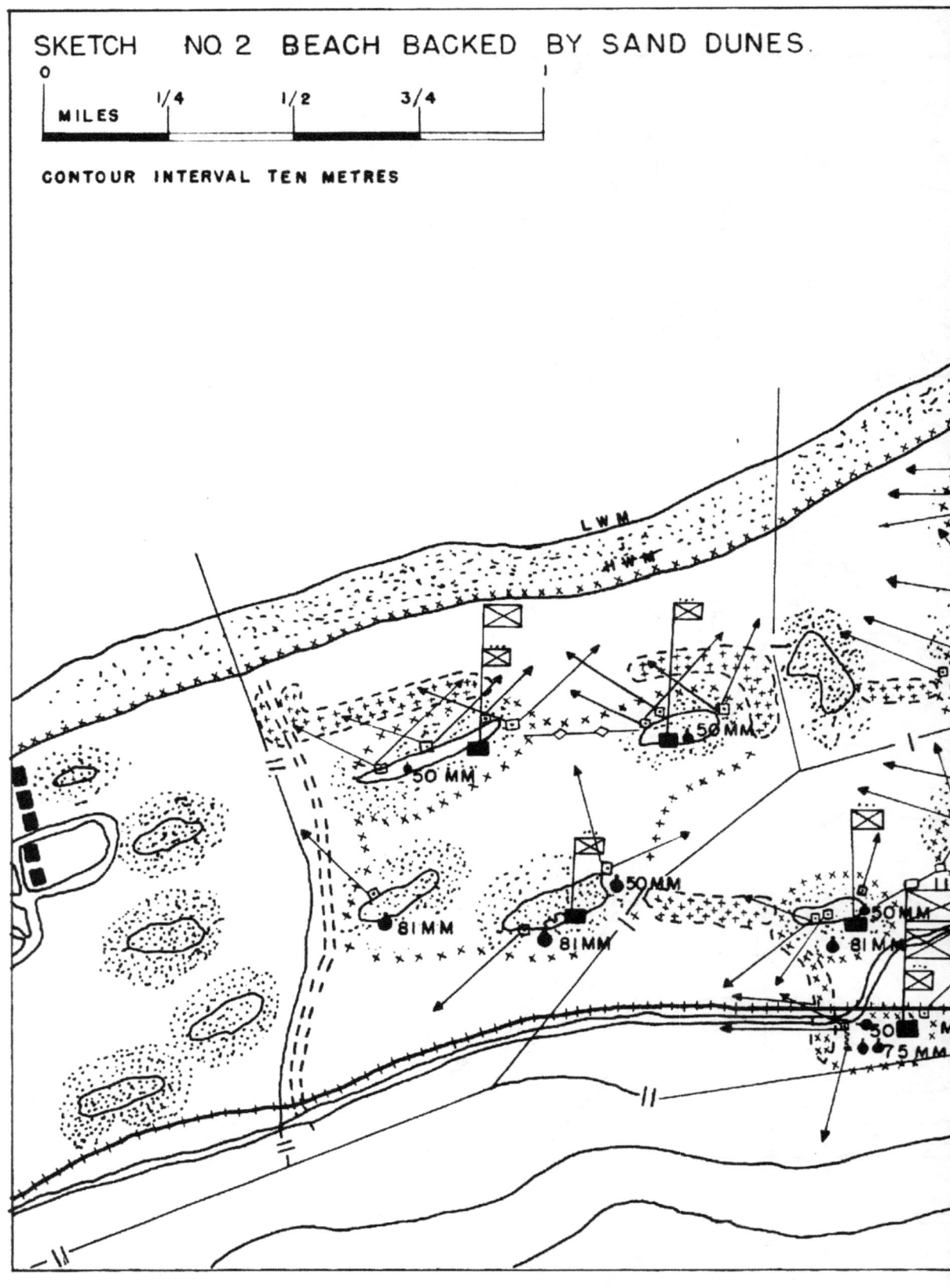

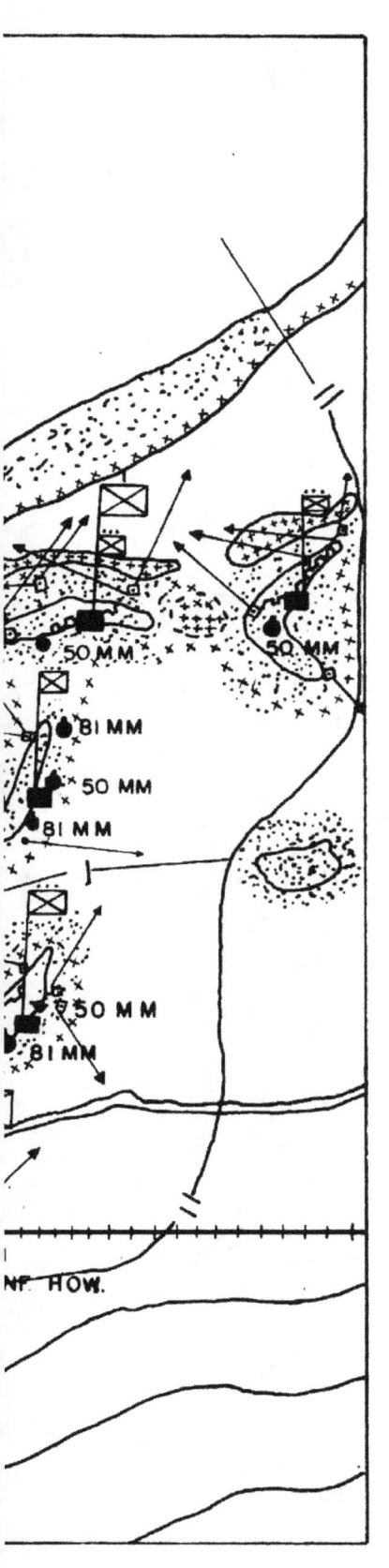

SKETCH NO.3 HIGH CLIFF WITH 2 SMALL STREAM EXITS

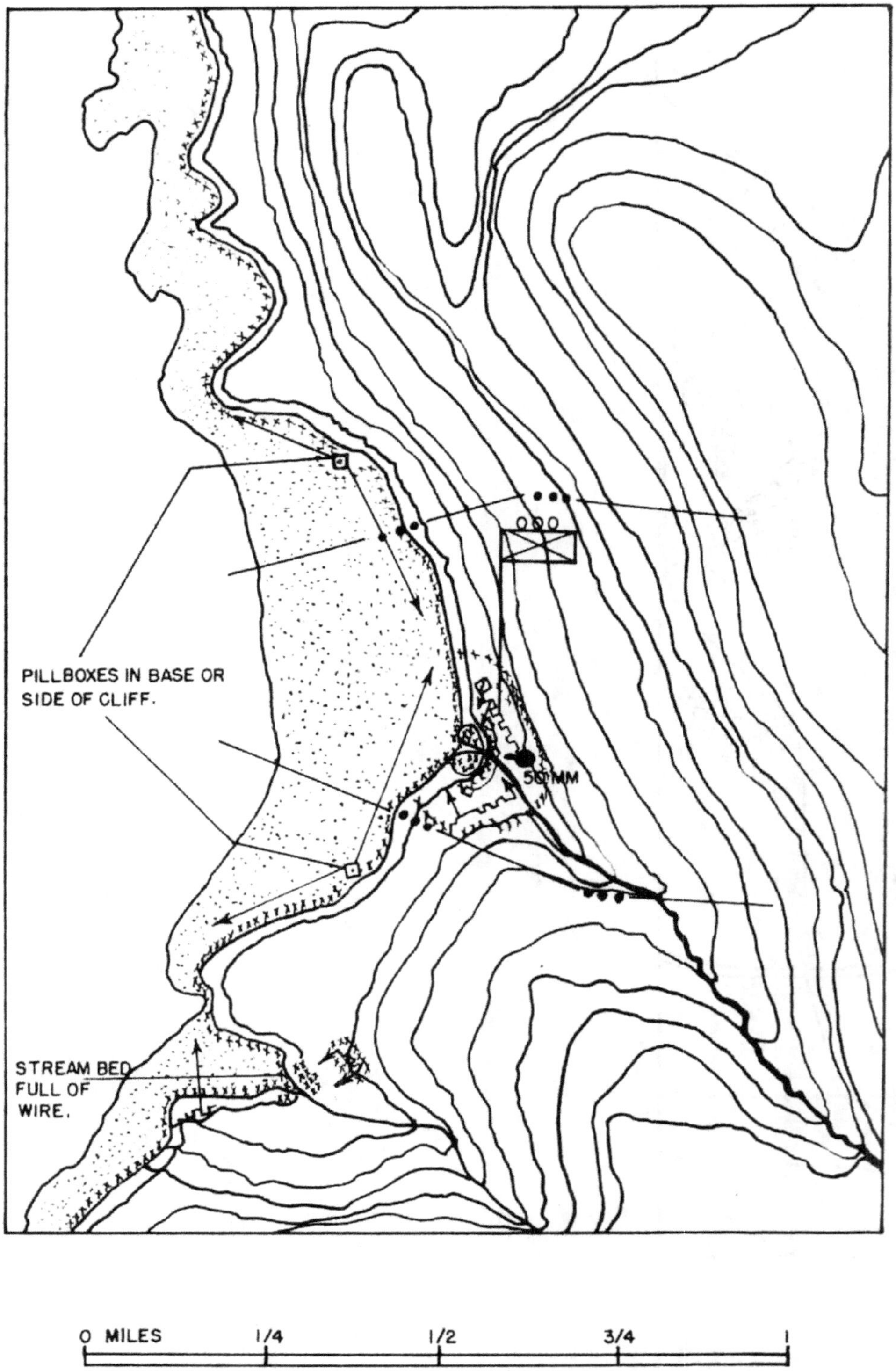

PILLBOXES IN BASE OR SIDE OF CLIFF.

50MM

STREAM BED FULL OF WIRE.

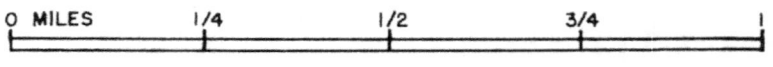

0 MILES   1/4   1/2   3/4   1

CONTOUR INTERVAL TWENTY METRES

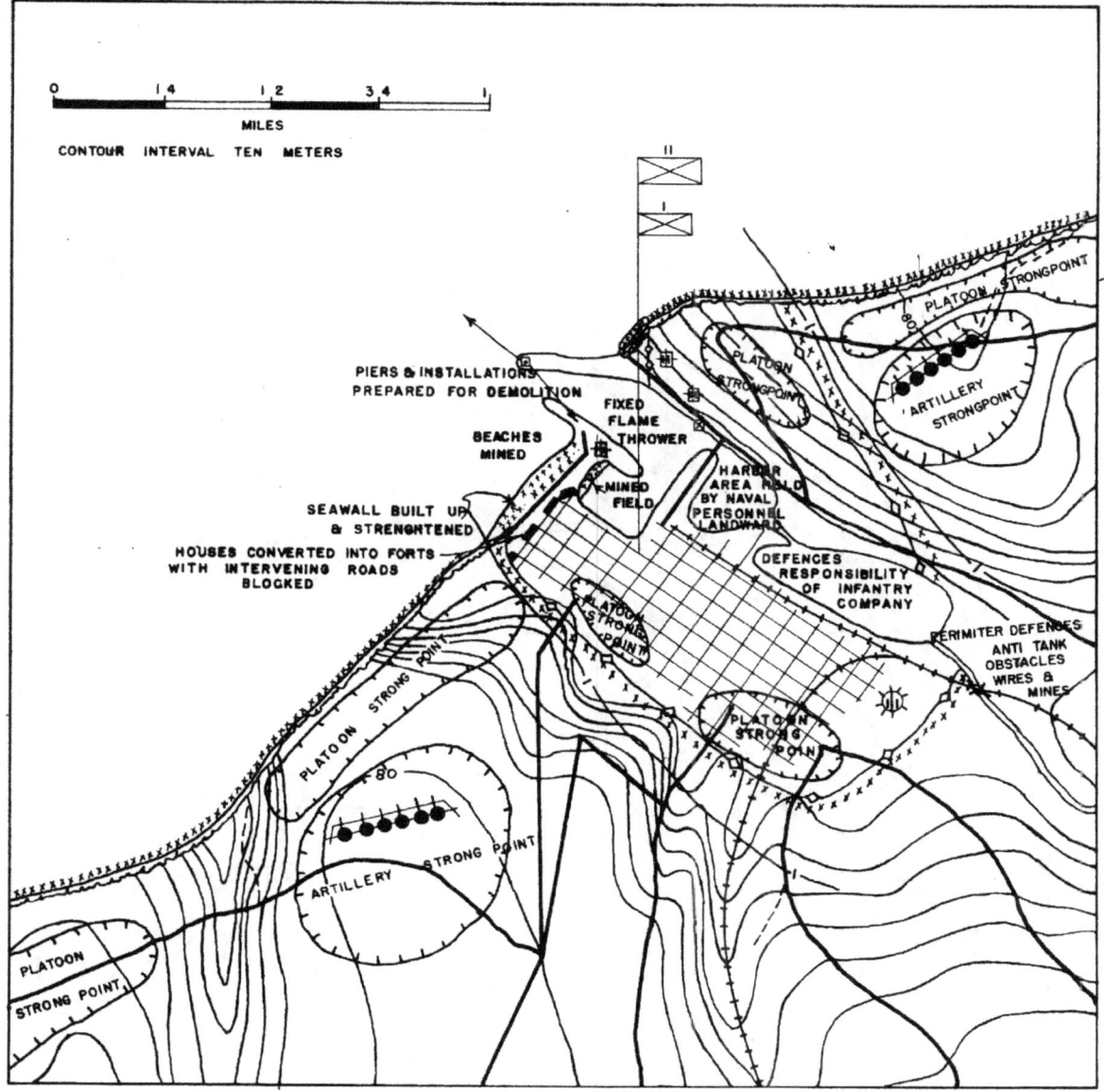

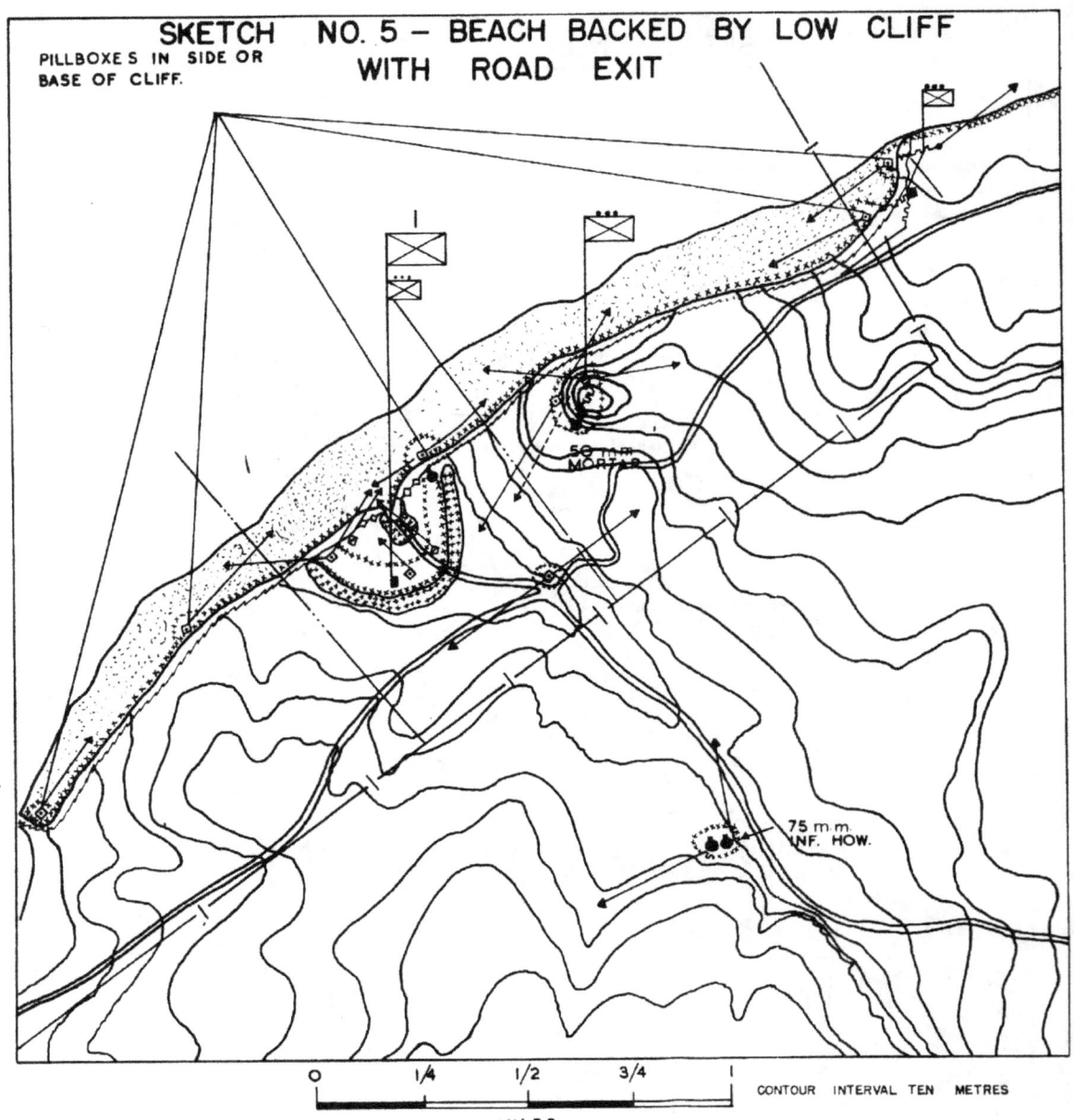

2.) **Beach Obstacles**: a.) <u>Wire</u>: In general, there are two or more continuous belts of wire along all open beaches, usually sited between high water mark and the back of the beach (photographs 1 and 2). In addition, all strong points are wired, belts varying in depth between thirty yards and two hundred yards according to the importance of the strong point. Beach exits, particularly gulleys in cliffs, are often barred by a dense wire entanglement. Wire is also extensively used in conjunction with minefields, walls, road blocks and other kinds of obstacles (photograph 3).

b.) <u>Mines</u>: The use of antipersonnel and antitank mines is general. The former are normally found round the perimeter of defended localities and at infantry exits from beaches; the latter at AFV exits from beaches, in open country behind beaches and at road junctions and road defiles. Anti-personnel mines also are scattered through antitank minefields.

c.) <u>Walls</u>: Of the various types of obstacles employed to prevent vehicles from moving off the beaches, the reinforced concrete or brick wall is the commonest type (photograph 4). Walls block streets leading from a beach or harbor, and also well defined exits from open beaches. Wall blocks together with existing buildings often form a continuous obstacle along the entire seafront of a town.

d.) <u>Antitank Ditches</u>: Antitank ditches are sometimes found surrounding strong points and on the seaward side of antitank or sea walls.

e.) <u>Steel Fence</u>: Steel fence obstacles are fairly common along stretches of open beach (photograph 5).

f.) <u>Others</u>: Dragons teeth, concrete pillars, rail pyramids, knife rests and inundations are also used.

3.) <u>Defenses</u>: a.) Pillboxes and shelters flush with, or nearly flush with, the ground and constructed of reinforced concrete at least three feet and sometimes six feet thick abound. They are sited so that fire from them can cover the obstacles. Pillboxes within strong points are linked up by highly developed trench or tunnel systems. The lower floors of houses along the seafront of coastal towns have also been walled up to form strong points (photograph 6). Gun positions are found in these houses, in antitank or sea walls and dug into the base of cliffs.

b.) <u>Defense layouts</u>: There are presented, as sketches 1 to 5, German defense layouts for various types of coast.

(1) <u>SKETCH 1</u>: This represents a beach backed by low cliffs, which may be scaled by infantry but are not passable for tanks or M.T. It is typical of much of the coastline between Calais and Le Havre. Points to note are:

(a) Pillboxes and emplacements are strung out fairly evenly along the cliff line. Pillboxes may be actually in the face of the cliffs.

(b) The heavy artillery battery has sufficient small arms for its own defense.

(2) <u>SKETCH 2</u>: This represents a battalion sector among sand dunes. Defenses which would be outside the battalion

sector are NOT plotted. The dunes can be crossed by tank (at least in between the dunes) and of course by infantry at all points. M.T. can also move between the dunes. A large part of the coast between the mouth of the Seine and the northeast tip of the Cherbourg Peninsula is like this. Points to note are:

(a) Pillboxes will be right in the sand dunes with only fire slits visible.

(b) Passages connecting pillboxes may be trenches or tunnels.

(c) Individual dunes are converted into strongpoints. Between the dunes extensive obstacles (particularly mines) are to be found.

(d) There is a great deal of wire used.

(3) SKETCH 3: This represents a cliff sector of coast, with two small streams running through it. The cliffs are not normally assailable. Both the streams provide beach exits for infantry. The larger one can take tracked vehicles in single file. This sector is typical of much of the coast of Britany. Points to note are:

(a) The thinness of the garrison, apart from the platoon at the main stream mouth.

(b) The concentration of wire in the smaller stream, covered by a few light machine guns.

(c) The antitank obstacles, minefield and wire in the larger stream.

(4) SKETCH 4: This represents the defense of a small port, such as may be found anywhere along the Channel Coast. Points to note are:

(a) The main responsibility for the defense of the port proper falls on the naval personnel, while the landward defenses are held by the infantry company.

(b) The town is defended on all sides by an antitank obstacle.

(c) The seawall has been built up, wired and reinforced as an antitank obstacle by a ditch dug in front of it.

(d) The houses facing the sea front have been turned into fortifications by bricking up the ground floors.

(e) All roads into the town, both from the sea front and from the rear, have been blocked.

(f) There are two coast defense batteries allotted for the defense of the town, to fire on shipping approaching the port, and three light dual purpose antiaircraft batteries located so as to be able to fire on landing craft.

(5) SKETCH 5: This represents a stretch of beach backed by low cliffs with a road exit from the beach. This may be found at many points along the coast. Points to be noted are:

(a) The concentration of strength and complexity of the obstacles around the beach exit.

(b) The two infantry howitzers which are allotted to the company to cover the beach exit.

(c) The normal pillboxes and emplacements strung along the cliffs (See also Sketch 1).

III. UNUSUAL CHARACTERISTICS

1.) <u>Information</u>: Information as to hostile dispositions will be obtained primarily from aerial reconnaissance and agents, and it must be accepted that such information at best will be incomplete. The result is that the planning for any operation against this area, for which the most complete information is desirable, must usually be done on less information than would normally be available. Therefore, trained reserves with special equipment to deal with the unexpected and unforeseen must be provided.

2.) <u>Limitation of Suitable Landing Beaches</u>: Due to the recognized limitation in the number of suitable landing beaches in any given locality and the difficulty of navigational control, the approach for the assault battalions over water must be executed as provided in the plan, or disastrous confusion is likely to result.

3.) <u>Absence of natural cover and concealment</u>: Other than in darkness, smoke or conditions of low visibility, troops coming ashore have no means of natural cover or concealment and are especially vulnerable to hostile fire. The absence of natural cover may be partially compensated for by armored craft and the dispersion of craft.

4.) <u>Lack of communications and control</u>: As the troops are being brought ashore, the only means of communication available are radio and visual. As it is impracticable to equip all craft with two-way radio, and as visual means can be used only within certain limits, control of craft and troops during this phase is at a minimum, with the result that any effort to divert craft and troops from the approach, as planned, is virtually impossible.

5.) <u>Absence of Normal Field Artillery Support</u>: The fire support in the initial stages of the attack is characterized by a lack of normal field artillery fire. Naval gunfire, air support, and fire from special craft and weapons currently classified as Secret must be substituted.

6.) <u>Uncertainty of landing at proper beaches</u>: Satisfactory equipment for guiding landing craft to the desired beach is still undergoing experimentation and development. Until a reliable method is assured, commanders must face the possibility of the failure to land at a planned location and must be prepared to exercise the utmost in initiative in adjusting to and overcoming the situation in a landing at an unexpected location.

7.) <u>Physical deterioration of troops in landing craft</u>: Plans should require troops to be in small landing craft for as brief a period of time as the situation will permit, as craft are crowded and men must sit in cramped and uncomfortable positions. Due to this reason, as well as to exposure to spray, cold and sea sickness, the solider's efficiency is progressively impaired and he is unable to fight at his best when the need is greatest, immediately upon reaching the shore.

8.) __Underwater Obstacles__: Defended beaches may be expected to include underwater obstacles which may take the form of piling, pipe scaffolding, mines, fixed flame throwers emplaced in the water, assorted debris, etc. Personnel with special equipment and special training must be provided to breach these obstacles. Military and naval personnel have a joint responsibility in this task which must be worked out and clearly defined, in advance.

9.) __Initial lack of normal transportation facilities__: During the early assault, the landing of motor vehicles is difficult and vehicles will be available only in small numbers. As a consequence, tactical planning must make provision for the hand-carrying of necessary equipment, ammunition and other supplies.

10.) __Support from the air__: Support for the attack from the air will consist of:

    a.) Antisurface vessel and antisubmarine patrol;
    b.) Fighter escort and cover;
    c.) Bombing;
    d.) Straffing;
    e.) Laying of smoke;
    f.) Reconnaissance

To effect direct support, an air support party will be attached to the headquarters of ground units, usually not lower than those of combat teams, to transmit requests of the Assault Echelon Commander, direct to the carriers or land bases. Direct support aircraft and air support parties will be on the same frequency, so that close liaison may be maintained.

11.) __Smoke__: To compensate for the absence of natural concealment in such water-borne movements, the use of smoke may be invaluable and should be carefully considered in planning. Due to variable wind conditions which may be encountered, alternate plans of employment of smoke must be prepared. These smoke plans require careful coordination with the plans for naval gunfire and air support. (See Smoke Annex).

12.) __Speed in execution__: The final success of the attack depends upon the attacker being able to gain and to secure essential terrain before the defender has time to block this effort by reserves and counter-attack. Attacking troops, therefore, must conduct their operations at a speed greater than that with which the enemy can effectively move and employ his reserves. The imperative need for speed at every step is regarded as the outstanding characteristic of an assault landing. Every officer and enlisted man must be taught, and must never be permitted to forget, that the degree of speed in every phase of an assault landing holds in balance the success of the entire operation.

IV. __WEAPONS, MEANS AND OBSTACLES.__

1.) __General__.

    a.) A combined operation, whether it involves a short sea voyage or a long sea voyage, presents certain definite problems which clearly differentiate it from any normal land operation. If the combined operation is to be successful, these problems must be carefully considered and solved during the planning stage. In their order of occurrence, they might be summed up as follows:

(1) Air and sea attack, while enroute;
(2) Reduction or neutralization of enemy coast batteries and Radar equipment;
(3) Close fire support on the beaches;
(4) Underwater obstacles;
(5) Reduction of obstacles on the beaches, to include possible cratering of the beaches, to provide a certain amount of cover for assaulting troops;
(6) Fire support during the advance inland.

b.) <u>The means</u>, with which to deal with the problems outlined above, are hereafter discussed in the broader categories showing the <u>possibilities and limitations</u> of each service involved.

2.) <u>Air Aspect</u>.

a.) <u>Air Operations in an Assault Landing</u>. In general, the following procedure and sequence of activity will be employed by the Air Force.

(1) An attempt to neutralize the enemy air actions that could be brought to bear against an assault landing in any particular location would first be dealt with by the strategic bombing and fighter forces. This will culminate in the selection just before and during D-day of certain airdromes, R.D.F. stations and control units for enemy aircraft.

(2) The next effort of protecting our operations from air and sea attack will be the air cover and escort, as performed by fighters, anti-surface vessels, and anti-submarine type aircraft, in their respective roles.

(3) In bringing the sequence of events to the actual landing, the next considerations deal with various types of enemy installations as objectives for air attack

(a) Airdromes within striking distance of the beachhead are an important objective, even though previous attacks have been made on them, as much neutralization as possible can and must be maintained.

(b) Enemy batteries for coast defense and anti-aircraft as well as other emplaced guns are important objectives but require direct hits within the emplacement for good effect. Destruction is most unlikely but short periods of neutralization may be effected by shock from H.E. bombs and smoke to reduce accuracy of fire. The types of action to accomplish this are: high, medium and low level bombing; dive-bombing; machine-gun and cannon strafing and smoke-laying aircraft.

(c) Enemy batteries not emplaced, such as field guns, railroad guns, artillery concentration of antitank guns, machine guns and other type weapons will be less difficult to deal with than emplaced weapons. In case of weapons of smaller caliber, such as the mortars, the areas in which these are located should consitute the objective for which the pilot is briefed, and the fire power will normally be delivered against the terrain feature, rather than the actual weapons.

(d) Fortified positions, with overhead cover, will vary in size from "hedgehogs" to individual pillboxes and these installations are, of course, made particularly for defense against air attack. Attack on them will require much combined study by both air and ground. As a rule, individual pillboxes offer almost an impossible target. "Hedgehogs" can be at least partially neutralized by the shock effect of heavy bombs (1000 pounds and over) and by smoke.

- 6 -

(e) Pillboxes, except perhaps a few that are located in very prominent positions, such as at the end of a pier or on a prominent landmark, will be almost impossible to single out accurately enough from the air to warrant an individual attack by aircraft. They will have to be included in the ground attack "bands" or "zones", that are attacked according to the prearranged plan.

(f) A type of objective which must be considered suitable for air attack, regardless of the difficulties which may have to be encountered, is the enemy RDF installations. These installations will generally be located on prominent high points and their exact location must be the subject of intensive advance study by our photo reconnaissance and intelligence agencies. Due to the fact that these installations must have aerials and a considerable amount of very sensitive equipment, they will be susceptible to both the shock action of heavy bombs and to the fragmentation effect of antipersonnel bombs. They are, of course, unaffected by smoke. In part, the same applies to signal centers; and to fire control centers and directors, although the location of these will not be as easy to determine. The RDF and similar installations must be attacked at the earliest practicable time, in order to obtain the maximum neutralization before the air battle of D-day.

(g) Another type of objective which will be attacked from the air for the purpose of obtaining some neutralization is the position of the defending ground forces which are without overhead cover; but which cannot be adequately dealt with until field artillery has been landed. These will be so disposed as to have a field of fire over the beaches and exits from the beaches, or will be in reserve positions, defiladed from naval gunfire. These objectives are highly susceptible to attack by fragmentation bombs, cannon fire and ground-strafing. They often cannot, however, be singled out by the pilot, even when these objectives have been located. Therefore, he will seldom attack specific small targets, attacking rather a terrain feature, and the aircraft thus generally must be furnished with the maximum number of anti-personnel bombs to create the best "pattern" effect. The attack on these objectives is accomplished by pattern-bombing, either in formation or "in trail", and by ground strafing. Some heavier bombs (100-300 pounds) may also be included, to obtain "shock" or "blast" effect. A study of the terrain covering the beaches, together with careful photo interpretation, will disclose the position areas in which these objectives may be found.

(h) Movement of large numbers of reserves, especially in vehicles, is a very suitable and profitable objective for air attack. The enemy must and will make movements of his reserves, soon after H-hour, which will continue until he has defined our complete intentions. These attacks on reserves cannot usually be planned in advance, beyond indicating principal routes which are available to the enemy. For this reason, and due to the extreme importance of blocking these movements, as much as fifty percent (50%) of the air formations, which have been briefed for attack of ground "zones", would be directed to attack large troop movements and their routes of march whenever seen, regardless of the mission for which they were briefed, provided that these movements are beyond the bomb line.

(i) Other enemy installations which the assault will encounter are wire and mine belts, and tank walls or obstacles. Here, the air can be of little help. The air attack of the hostile position areas or ground attack zones will cause some incidental damage to surrounding wire or mine belts, but its effectiveness must be regarded as limited. The breaching of tank walls or obstacles by air attack is regarded as wholly impracticable and these walls and obstacles cannot be considered as suitable targets. The air can, however, attack personnel and equipment using these walls or obstacles for cover.

    b.) <u>Airborne Troops</u>

        (1) In selecting appropriate missions for airborne troops, the tendency to dissipate their strength by scattered employment should be guarded against; these troops are lightly armed and should be used in mass to be most effective.

        (2) Airborne troops should be directed against objectives which are vital to the force as a whole, but only against objectives which cannot be reasonably attacked by other ground forces.

        (3) Inasmuch as the hostile reaction to an assault landing may likely take the form of armored counter-attack, large airborne units should not be landed so deep within the enemy area as to preclude reasonable prospect of their being given timely reinforcement by other ground forces.

        (4) In assisting an assault landing, the following are considered appropriate missions for airborne troops:

           (a) To attack key points in, or in the rear of coastal fortifications, such as gun positions inland from the beach;

           (b) To block movement of enemy reserves;

           (c) To assist in securing objectives of the assault division;

           (d) To attack beach defenses from the rear;

           (e) To disrupt enemy communications and supply facilities.

        (5) Airborne units as large as a division may appropriately be used to engage hostile reserves, and to assist in securing the objective of the assault division. They may assist in blocking counter-attacks of large enemy forces, but acting alone are incapable of holding hostile armor.

3.) <u>Rangers</u>

    a.) The organization and training of Ranger battalions make them ideal units to be employed in the early capture and destruction of coastal guns and other installations, the early reduction of which is vital to secure the flanks of an assaulting force. Such troops are trained to attack in unlikely places and under unlikely conditions to gain surprise. They could also be usefully employed for feints and taking of objectives from the flank and rear, by landing upon beaches that would not be suitable for landing of usual assault formations.

Revised 9 July

4.) Fire Support.

   a.) Naval Bombardment.

      (1) Naval gunfire may be expected effectively to neutralize, during the period of firing, areas of a dimension suitable to the caliber of the guns. High capacity (HC) ammunition and airburst AA ammunition have increased the effect over that obtained in the past. Harassing and interdiction may be included as types of effective neutralization by naval gunfire.

      (2) Naval gunfire is less efficient for fires of destruction. Good results, however, may be expected when the target is imposed and visible from the firing ship and when the range is not too great. If hits can be obtained, the effect of large caliber AP shells against concrete emplacements should be satisfactory.

      (3) Each ship may be considered as composed of a number of artillery batteries, equal to the number of its fire controls. A single ship can thus engage simultaneously two or more targets and, in the case of larger ships, with different calibers and types of shells.

      (4) It is reasonable to expect that combatant ships (BB, CA, CL and DD) will be furnished to provide gunfire support for a landing operation on the following scales: one ship for the direct support of each assault battalion.

      (5) Within the limits of ammunition supply, naval gunfire may be expected to provide:

         (a) Preparation in advance of the landing;
         (b) Neutralization of areas containing enemy installations directly opposing the initial troop landings;
         (c) Neutralization of selected rear areas (CP's, OP's communication centers, concentrations of reserves);
         (d) Counter-battery, and neutralization of both field artillery and coast artillery;
         (e) Fires on targets of opportunity designated by the supported troops.

      (6) Fire will be as effective as the personnel who control it are competent. Two specially trained types of personnel are required in addition to normal ships complement:

         (a) Naval officers for shore fire control and liaison with troops.

         (b) Army officer on each firing ship to translate troop requirements into terms of naval gunfire support.
         (Each of these must be provided in addition to regular T/O and must be thoroughly trained in advance of an operation.)

   b.) Fire Support on the beaches, to include cratering of the beaches.

      (1) Employment of field artillery in the early stages of a landing operation differs from its employment in ordinary land operations in the following essential features:

(a) In ordinary land operations, from the beginning field artillery executes preparatory and supporting fires. In a landing operation, preparatory and supporting fires are executed by naval guns until field artillery is ashore and prepared to reinforce the ship fire or to take over certain fire missions.

(b) Due to difficulties in transporting and landing guns and ammunition, the amount of field artillery available in a landing operation is usually less than that in a land operation on a corresponding scale. This factor may require ship guns to continue on certain fire missions during all or a large part of the operation. Coordination of fire of field artillery and ship guns is required.

(c) Normally, in a landing operation, field artillery must reach the beach before it can go into action. This factor, together with the necessity of reinforcing or relieving naval guns at the earliest possible time, makes it necessary to employ field artillery with great boldness.

(d) In offensive operations on land, field artillery coordinated its fire with the advance of the infantry from the beginning of the attack. In a landing operation, the field artillery support begins after the attack is well under way. Liaison with the front line troops is continuous in order that close supporting fires may be delivered as soon as batteries are prepared to fire.

(e) Because of the impracticability of exercising centralized control, field artillery batteries are usually attached to infantry units during the initial phases of a landing operation. This makes difficult the concentration of fire of a number of batteries on a designated objective. As soon as the situation on shore permits, field artillery is placed under centralized control.

(2) To meet this lack of fire support from organic artillery, certain means have been developed and others are now in the process of development.

(3) <u>Cratering of the Beaches</u>. Cratering of the area over which the assault must pass is essential to provide cover for the assault troops. Normally, this will occur incidental to aerial bombardment and shell fire on this area.

c.) <u>Special Craft</u>.

(1) Special types of craft constructed for close fire support of landing troops (such as LCS, LCG, LCF, LCT rocket) should be considered separately and apart from naval gunfire, since their capacities and employment are very distinct from those of regular combatant vessels.

(2) The essential feature of these craft is that they do not lie off-shore and deliver long range fire, but actually accompany landing craft into the beach firing while troops are getting ashore. These craft thus belong in the first wave of landing craft and become a specie of artillery, emplaced at the water's edge, delivering pointblank fire against targets which are in immediate opposition to the troops. (The rocket craft is an exception). This does not, of course, preclude fire while off-shore.

(3) Although a technique for their employment still remains to be developed, (in the U.S. service), two points may be noted as to their number and disposition.

(a) The numbers will depend on the number of guns necessary on particular beaches in order to provide fire support after the naval gunfire has lifted. This will be dependent on enemy installations and must be especially considered for each case.

(b) These craft must be disposed (on the flank, in the center) according to the enemy dispositions, so that each may place its designated target under fire, while the target is being attacked by the troops.

d.) <u>Organic artillery in Landing Craft</u>.

(1) There are certain developments in the nature of specialized use of landing craft which will materially increase the fire support to be expected from organic field artillery. In a short sea voyage, organic artillery may well be transported from shore to shore in LCTs. By replacing towed guns of the division artillery with 105mm self-propelled guns, it is possible to give some degree of fire support while these guns are still afloat. It is not considered feasible to conduct indirect fire from these craft. However, direct fire from ranges up to 10,000 yards have proved very effective.

(2) Following is the suggested organization and scale of equipment for a battalion of the organic field artillery of the infantry division to enable it to conduct this type of fire:

(a) 3 batteries of 6 guns per battery, armament 105mm;
(b) self-propelled guns to be loaded 3 into each LCT;
(c) 2 LCTs per battery, 6 LCTs per battalion

(3) Artillery, so equipped and transported across the water gap, will be able to land at a much earlier time than has been considered possible in the past. Organic artillery employed to fire from craft while afloat, must not become so involved in firing as to fail to get ashore at the earliest moment possible so as to assume its primary mission of close support to the infantry.

e.) <u>Artillery support during the initial stages of the advance inland</u>.

While it is possible, by the use of self-propelled guns in LCTs, in a shore to shore operation, to provide some artillery support at an earlier stage than heretofore, naval gunfire and air support will have to increase this support to the maximum extent possible.

5.) <u>Smoke</u> - (See Smoke Annex.)

6.) <u>Reduction of Obstacles</u>.

a.) <u>Underwater Obstacles</u>.

(1) As agreed by the Army and the Navy, the Navy is responsible for the removal or breaching of underwater obstacles of all types which exist to seaward of the normal grounding points

Revised 9 July   - 11 -

of the landing craft, at the time and place of landing, and the landing force (Army or Marine) is responsible from these points inland. Each service must be prepared, in mental attitudes as well as by training, to assist the other.

(2) The Navy has directed the establishment of naval demolition units to perform its share of this task. These are in the process of organization and of development of equipment and technique.

(3) Well in advance of any operation, these units should be provided for joint training with Army engineers in the theater of prospective activity.

b.) Obstacles on the beaches.

(1) The problem of breaching the defensive system on the enemy-held coast, the so-called "crust", resolves itself into two general classifications:

(a) Neutralization of fire covering the obstacles;
(b) Reduction or passage of the obstacles themselves.

c.) Fire covering the obstacles.

(1) A feature of the defensive system is a large amount of artillery and automatic weapons of all kinds employed for the defense of the coast. In addition to heavy and light coastal defence guns and mobile railway guns which are mainly in the vicinity of the ports, there are:

(a) Land batteries of medium caliber, fitted with instruments to engage ships at sea and, in some cases, also able to fire on the beaches.

(b) Antiaircraft and antitank guns sited where they can, if necessary, engage craft attempting to land;

(c) Field and medium artillery, sited to bring fire to bear on beaches where a successful landing has been made, but from such a distance that the guns themselves will not become involved in the fighting for the beaches. There are on the average some 200 guns of this nature employed to cover a divisional sector.

(2) All of the above weapons must be dealt with by means other than those immediately available to the assaulting forces. The assaulting forces will have, as their immediate problem, the removal of direct fire from the obstacles. This fire will emanate from antitank weapons and machine guns in concrete emplacements. These emplacements are of such a nature and so well concealed that long range naval bombardment and air bombardment will have difficulty in locating and hitting them in vulnerable spots. The weapons which would normally be used in land warfare to counteract this fire are too vulnerable to be placed upon the beaches in the initial stages of the assault.

(3) The two most effective means at present possible are, fire from LCS and LCGs, and fire from tanks which are brought ashore with the assault wave.

Revised 9 July

(4) It is important that these tanks should be not considered as tanks in the accepted sense. They will probably never leave the beach, due to mines and extremely heavy fire on the beaches. They become, for this purpose, armored assault guns, and their inclusion in the assault waves will be necessary to replace the fire of assault weapons normally available for the reduction of fortifications, using the technique set forth in T.C.-33.

d.) Reduction of Obstacle-Means.

(1) Hand placed charges: Hand placed explosive charges are the most reliable means of destroying concrete and steel obstacles. When operating under fire or when it is necessary to cross fields of small arms fire, a tank adapted for transporting engineers and explosives can be effectively used.

(2) Wirecutting parties: Wirecutting parties are composed of one or more soldiers equipped with hand wire cutters. They can operate either under protective fire or under cover of darkness or smoke. Individual strands of wire are cut and thrown back until a gap of the desired size is cleared.

(3) Mine removal parties: Hand methods, using mine detectors or probes, are the most reliable means of creating passages through minefields.

(4) Bangalore torpedoes: Bangalore torpedoes are tubes fitted with explosive. The present ones are about 2 inches in diameter and come in 5-foot sections, however, sections can be quickly joined together. The torpedo is pushed into place by hand and then detonated. They will effectively clear gaps in wire obstacles and will also detonate mines.

(5) Snakes: Snakes are essentially heavy Bangalore torpedoes, that are moved into position by tanks. A special nose on the front end permits the snake to be pushed over uneven ground. Snakes of an effective length of 100 yards have been used.

(6) Preparation fires: Heavy preliminary bombardment from the air and sea can be expected to effect sufficient destruction of minefields and wire obstacles to permit passages of infantry, but cannot be relied upon to provide gaps for vehicles. A very heavy concentration is required for even this result. Such a heavy concentration may be obtained by use of a special craft now being developed.

(7) Carpets: Carpets of chicken wire or other similar material can be placed over barbed wire either by hand or by special carpet laying devices mounted on Bren carriers or tanks. These carpets will permit the passage of infantry over the wire.

(8) Fascines: Fascines carried in a special rack in front of a tank can be dumped in anti-tank ditches or against low walls. Tanks supported by these fascines can then ride over the obstacle. This is a rapid method of crossing anti-tank ditches and low walls, however, due to its unusual load, if seen, the fascine tank will undoubtedly draw hostile fire.

(9) <u>Grapnel and Cable</u>: Grapnels on cables can be pulled behind tanks to destroy wire obstacles. Also light grapnels on cables have been projected from tanks by means of rockets. The chief use for these is pulling trip wires in minefield.

(10 <u>Mechanical devices</u>: There are being developed various mechanical means for clearing minefields, destroying wire and breaching obstacles. As these are proven practical and are made available for issue they should be incorporated in planning and training.

V. Organization

1. Basic Units

   a. Normal type rifle platoons are reorganized into two types of special platoons for the assault of a fortified beach - Assault Platoons and Support Platoons.

   b. The Assault Platoon is a specially organized team, trained to locate and quickly breach and reduce defensive works of a fortified position at most favorable points in the platoon zone of action, in order to permit other assault troops to advance rapidly without undue losses.

   c. The Support Platoon provides fire support for the Assault Platoons.

   d. There are two Assault Platoons and one Support Platoon in each Assault Company. However, the number of platoons may be adjusted to the situation.

2. Assault Platoon

   a. The Assault Platoon consists of three (3) sections - a Reconnaissance Section, a Support Section and an Assault Section, performing respectively, the functions of reconnaissance, firing and assault, i.e., breaching and reducing hostile defensive works. The sections of the assault platoon are composed as follows:

   (1) Reconnaissance Section:

   SCOUT PARTY

Personnel	Arms & Equipment
3 EM and 1 Officer or NCO	Carbines or pistols, grenades, wire cutters, compass.

   OBSTACLE CROSSING PARTY

Personnel	Arms & Equipment
4 EM and 1 NCO (Sgt)	Carbines, pistols, grenades, wire cutters, compass, bangalore torpedoes and wire mat for crossing wire entanglements

   (2) Support Section

Personnel	Arms & Equipment
1 Leader NCO (Sgt)	M-1 Rifle
1 Assistant Group Leader	M-1 Rifle
1 Automatic Rifleman	Automatic Rifle
1 Asst. Automatic Rifleman	Carbine
1 EM	Submachine gun
5 EM	M-1 Rifles

(3) <u>Assault Section</u>

Personnel

1 Group Leader (off)
1 Assistant Leader
4 Rocket (or grenade)men
2 Flame Thrower operators
3 Demolition men
5 EM (Wire cutters & wire mat carriers)

Arms & Equipment

Flame Throwers, Demolition charges, Rocket launchers (or grenade dischargers), Bangalore Torpedoes, wire cutters.

Personnel

Arms & Equipment

Wire cutting pliers, Signal projector, carbines, smoke grenades.

 b. In addition to the arms and equipment indicated above, the following may be carried:

   Intrenching tools
   Wire cutters
   Hand grenades (offensive)
   Smoke grenades
   Compass

3. <u>Support Platoon</u>

 The Support Platoon is a normal type infantry rifle platoon with two (2) 60 mm mortar and two (2) LMG teams attached. The mortars and LMG's will frequently be attached to Support Sections of the Assault Platoon for the initial landing phase.

 The Support Platoon, in addition to giving fire support to each Assault Platoon, may also be used to deepen penetration in the zone of action of either Assault Section.

4. <u>The Assault Company</u>

 The Assault Company (See Chart B) is organized into a Company Headquarters, two Assault Platoons and one Support Platoon. It is designed to permit the Company to attack on a two platoon front. However, the depth of the obstacle in front of the defensive position may require the successive operation of the Assault Sections. In such case, the two Assault Platoons should be employed in column and both Assault Platoons could be landed at the same beach. Where the situation requires, the number of Assault Platoons may be changed as in the Example included in Section VI.

5. <u>The Assault Battalions</u>

 The Assault Battalion (See Chart C) is organized into a Battalion Headquarters and Headquarters Co, two (2) Assault Companies, one (1) Support Company and one (1) Reserve Weapons Company. This organization permits employment of the battalion on a two company front, with a Support Company having two of its platoons strong in fire power, which can be used: (a) to reinforce both Assault Companies, if companies are employed abreast, or (b) as one unit to exploit success in either assault Company zone of action. The Reserve Weapons Company is equipped with the surplus weapons of the Rifle Company Weapons Platoons and the Heavy Weapons Company.

These weapons provide the Battalion Commander with a reserve which he may use for replacing Assault Company weapons or for increasing fire power at any point in the battalion zone of action. One quarter ton (¼ T) trucks landed with the Reserve Weapons Company should correspondingly increase the effectiveness of the reserve fire power.

6. The Assault Regiment

The Assault Regiment is organized into a Headquarters, Headquarters Company, Service Company, Anti-tank Company, Anti-aircraft Platoon (cal. 50), Medical Detachment and three battalions.

Two of the battalions are organized into Assault Battalions, as described in par (a) above.

Each of two Assault Battalions provides the regiment with four Assault Platoons for breaching hostile defensive positions. This permits the regiment to land on a broad front and breach as many as eight (8) passages through the obstacles; or, should the obstacles have considerable depth, to concentrate a sustained drive on a narrower front.

The Reserve Battalion is a normal T/O organization (less many of its vehicles) and has the mission of exploiting the success of the Assault Battalions.

All unlettered organizations of the infantry regiment are organized and equipped in accordance with authorized tables.

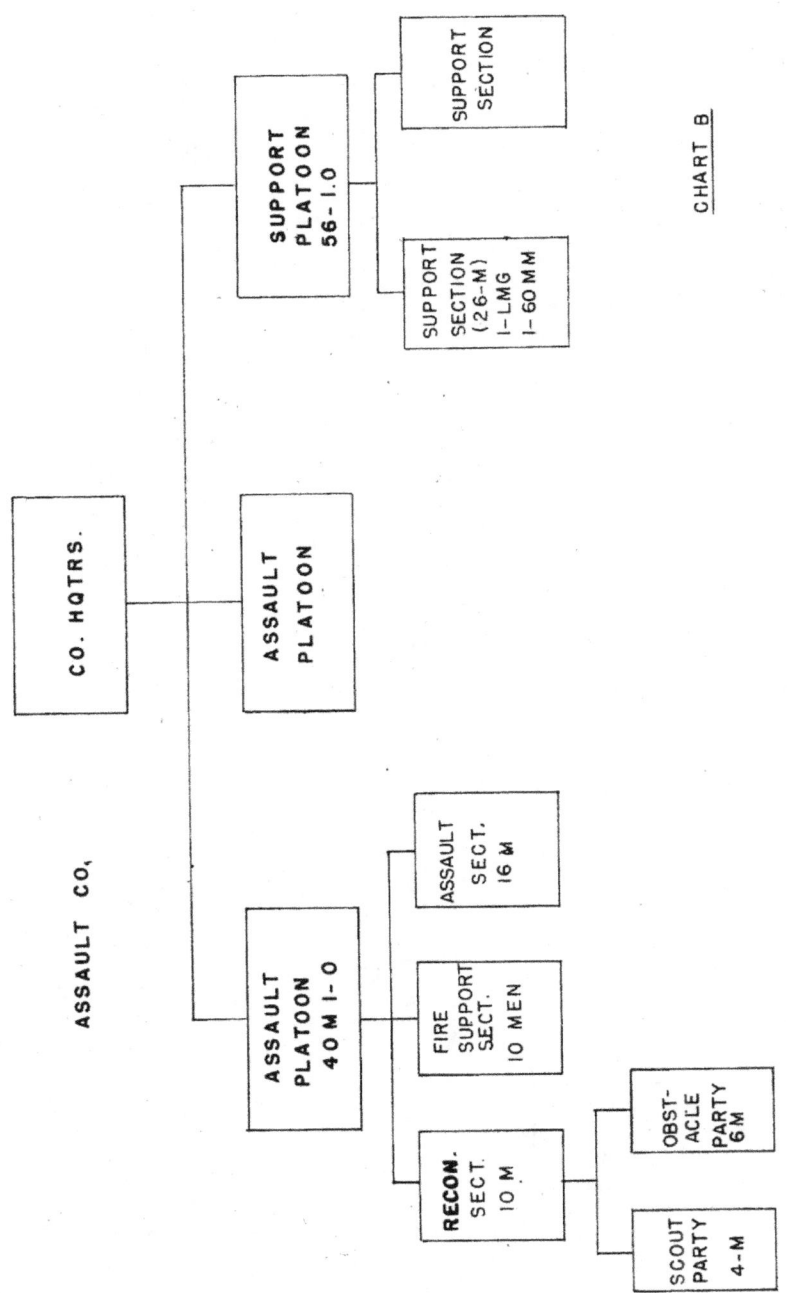

CHART B

AUTH. GRANTED C.E. MAP AND REPR.
SECT. HQ. S.O.S. ETOUSA TO REP.
O NEGS AND 215 POSITIVES.
SIGNED
RANK
SECTION
DATE 29 JUNE 1943

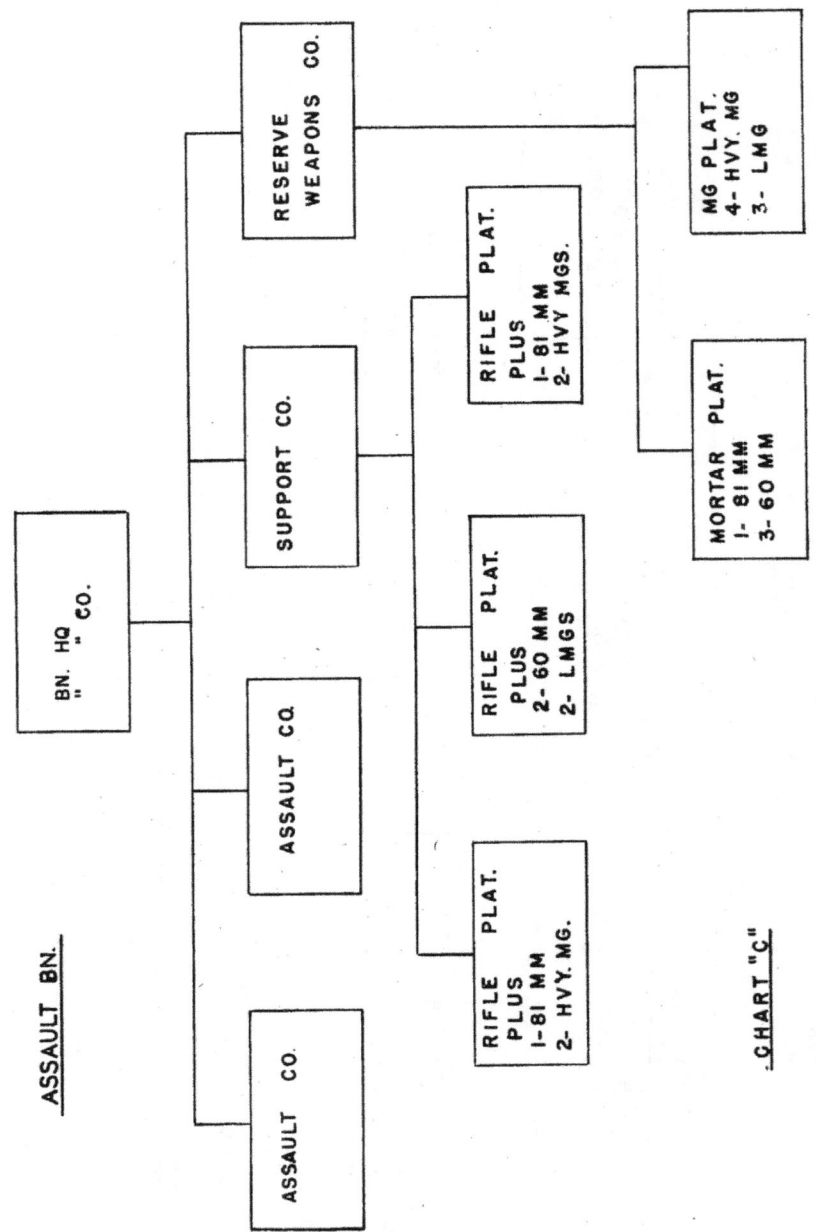

AUTH. GRANTED C.E. MAP AND REPRO.
SECT. HQ S.O.S. ETOUSA TO REP.
 NEGS AND 215 POSITIVES.
SIGNED
RANK
SECTION
DATE 29 JUNE 1943

ASSAULT TRAINING CENTER
CONFERENCE
HQ. ETOUSA

VI. EXECUTION OF THE ATTACK

1.) <u>Sequence of Operations</u>:

<u>a</u>.) In order successfully to attack and penetrate a strongly fortified beach area, it is essential that hostile defenses undergo a softening up by air bombardment for several weeks prior to the attack, supplemented by naval bombardment preparatory to the assault. Furthermore, at the time of, and during the attack, local air superiority must be obtained. Occasional single hostile planes may slip in, but attacks by formations of enemy planes must be prevented. Continuous air and naval fire support must be available on call during all phases of the attack.

<u>b</u>.) The chronological sequence of the phases of an attack might be as follows:

(1) Rangers land during darkness to destroy key beach defenses, inaccessible to naval gunfire and air bombardment, and difficult for assaulting troops to attack.
Example: Well-concealed heavy batteries located on rocky headlands.

(2) Parachute troops are dropped behind the defenses prior to nautical twilight to secure a base of fire for airborne troops, and to attack strongly defended localities and artillery located well inland.
Note: The landing of Rangers and the dropping of parachute troops must be coordinated so that neither precludes surprise by the other.

(3) Naval bombardment begins at nautical twilight.
Targets: Hostile batteries and strongpoints commanding the beach.

(4) Air bombardment begins as soon as there is sufficient light for accurate bombing.

Targets: Areas from the waterline inland to include batteries and strongpoints commanding the beach. It is important that minefields and wire be included and that the beach be left well cratered in order to provide cover for assaulting troops.

(5) A heavy smoke screen is laid by planes and maintained by planes and mortars from landing and support craft, to cover movement of troops into shore.

(6) Glider elements of airborne troops start arriving as soon as there is sufficient light to land gliders.

(7) Leading wave reaches the beach, assault platoons are landed, and tanks are beached or positioned hull down in the water.

(8) Obstacles are breached and assault platoons close with the pillboxes.

(9). Remainder of assault companies and the battalion reserve elements land, pass through the assault, the obstacles and push rapidly inland to secure the battalion objectives.

(10) Reserve battalions land on beaches where going is easiest, push inland to deepen the penetration, and secure the Regimental Intermediate Beachhead Line.

(11) Division reserve lands, moves rapidly inland and secures the key position of the Division Intermediate Beachhead Line.

2.) <u>Assault Principles</u>:

a.) The Force Commander must look well beyond the initial assault and consider the battle between his force and the hostile main reserves. Operations necessary to the deployment of his total force ashore must be characterized by utmost speed, for these operations are matched in a race with the enemy's movement of mobile reserves.

b.) The Assault Division will have the mission of seizing, defending and maintaining a beach head. The beach head, successfully defended, must:

(1) Secure suitable and adequate beaches and exits for passage of the follow up force.

(2) Provide space for offensive deployment of the buildup force.

(3) Have its perimeter within the defensive capabilities of a division.

(4) Have its perimeter on terrain favorable to the defense.

c.) The mission of the Assault Division may be considered in three phases:

(1) Seizure of beaches with suitable exits for passage of troops, artillery, tanks and other supporting weapons.

(2) Attainment of intermediate objectives which secure the landing beaches against small arms and other direct fire; and defense of these objectives against local counter-attack.

(3) Seizure of objectives on the beach head perimeter and defense of the beach head against counter-attack by hostile main reserves.

d.) Concurrently with these operations, improvement of the landing beaches and exits will be initiated and carried on for the logistical support of the division, and <u>in preparation for expeditous passage of the build up force</u>.

e.) Seizure of beaches with suitable exits may be accomplished by direct frontal attack, or by seizure of adjacent beaches which, due to lack of exits, are most lightly defended.

The latter would be followed by envelopment of the defenses covering the desired beach. Choice between these two schemes of maneuver can be made only on the basis of careful study of terrain and information of the disposition of enemy defensive installations. Initial attack through weakness, followed by envelopment of the desired beaches, is the better choice wherever possible. The balance between success and failure is never more delicate than in the initial assault; therefore this assault should be directed against the least strength. The advantages outweigh the disadvantages of foot movement and hand-carry of machine guns and mortars over difficult terrain.

f.) Normally a battalion may land on, and attack the defenses of about 500 yards of beach. The frontage suitable for a division will depend upon the number of, and space between, suitable landing beaches, and exits, the inland terrain, and defensive supporting strong points. As a rough rule a frontage of about 3 miles may be considered suitable.

The Regimental Intermediate Beach Head Line must secure the landing beach against direct fire and effective mortar fire. It should include terrain favorable for defense against counter-attack by local reserves. It must be within the capabilities of the regimental combat teams to seize and defend without assistance of the division reserve. As a rough guide, about 3,000 yards inland appears reasonable.

Considerations affecting the location of the Division Intermediate Beach Head Line, or beach head perimeter, are discussed in paragraph (b) 1.). As a guide, a distance of 5 to 6 miles is necessary to keep effective hostile medium artillery fire off the beaches.

3.) <u>Reduction of Fortifications</u>: a.) <u>General</u> - The first task confronting the assault battalion is to pierce beach defenses and to gain a toe-hold on terrain providing cover and concealment just beyond the sand beaches. This involves closing with and destroying concrete pillboxes covering the beaches and requires special assault platoons, (see section V), properly organized and equipped to cut through wire and other obstacles, to get through mine fields and to silence pillboxes by means of explosives, probably hand-placed.

b.) <u>Fire Plan</u> - Assault platoons can land, get through obstacles and close with pillboxes only if direct defensive fires, which can be brought to bear on them, are beaten down and finally neutralized during their approach, landing, and attack, by an integrated, progressive scheme of fires, employing <u>all</u> means. Smoke is equally essential and must be carefully coordinated with the fire plan and the attack, so as not to mask fire targets or confuse the assault platoons. Shore batteries covering approach of landing craft must be silenced. This can best be accomplished by naval gunfire, air bombardment, and assault by Rangers, the latter landing and reaching the batteries by stealth under cover of darkness.

(1) The first phase of the fire plan is the general softening by air bombardment, beginning several weeks in advance, building up to the maximum level that can be maintained, and maintained at that level until D-day. During the assault, air bombardment concentrates on inland areas containing defenses most dangerous to the assault.

(2) The second phase is the naval gunfire phase, beginning with bombardment of shore batteries and beach defenses on a planned scheme of fires, and lifting with the approach of assaulting troops, to inland hostile fire support emplacements;

and/or shifting laterally to beach defenses flanking the landing beaches.

(3) The third, or direct fire phase, begins before assault waves come within direct fire of beach defenses, and consists of direct fire on enemy embrasures delivered from support craft, artillery on landing craft and tanks. (Close liaison with current study and development in the field of direct fire support for assault waves must be maintained in order to capitalize on all proven developments).

(4) The choice between blinding hostile direct fire by smoke or neutralizing it by direct fire requires careful study of known information of enemy defensive installations, terrain and weather. A smoke cloud once laid is controlled only by atmosphere and terrain, while control of gunfire is positive. Obviously, direct gunfire cannot be laid on targets which are in smoke. Probably both smoke and direct fire will be used to complement each other in a coordinated plan.

c.) <u>Assault Platoons</u> (1) - The Assault Platoon should be carried in one craft of suitable capacity, and should be loaded so that the Reconnaissance Section is the first to go ashore. See charts A and 1-A.

(2) The Platoon, accompanied by supporting tanks, lands under the protection afforded by support fire and smoke.

(3) The tanks, from positions on the beach hull down in water, deliver direct observed fires on embrasures from which destructive fires would otherwise be brought to bear on the Assault Platoons. These fires continue until the Assault Platoon is prepared to close with the pillboxes or other defenses and demolish them. Such support must be closely coordinated.

(4) The Scout Party of the Reconnaissance Section, operating in accordance with the accepted principles of scouting and patrolling, rapidly locates the hostile weapons and positions and selects the least difficult places for passage through an obstacle.

The Obstacle Crossing Party, taking advantage of all available cover, moves up to the obstacle at the points selected by the Scout Party and prepares to cross it or breach it. Wire mats or bangalore torpedoes may be used for this purpose.

(5) The Fire Support Section, with 60 mm mortar, and light machine gun teams attached, is landing immediately behind the Reconnaissance Section. As it debarks, the section should move quickly in suitable deployed formation to selected firing positions from which it may support the action of the Assault Section. Their positions may well be on the flanks of the gap through which the Assault Section will cross the obstacle.

(6) The Assault Section is the main force of the Assault. Consequently, on landing, care must be exercised to prevent its unnecessary exposure to hostile fire. It should immediately take up a suitable deployed formation and take momentary advantage of whatever cover is available while preparations are made for its employment. The mission of the Assault Section is to knock out defensive works which threaten to hold up the advance of assaulting troops, and it will utilize the assault principles set forth in I.C. No. 33, "Attack of Fortified Position". It operates under close support from the Support Section and the Company Support Platoon as well as accompanying tanks. Therefore, steps must be taken to coordinate the actions of these units.

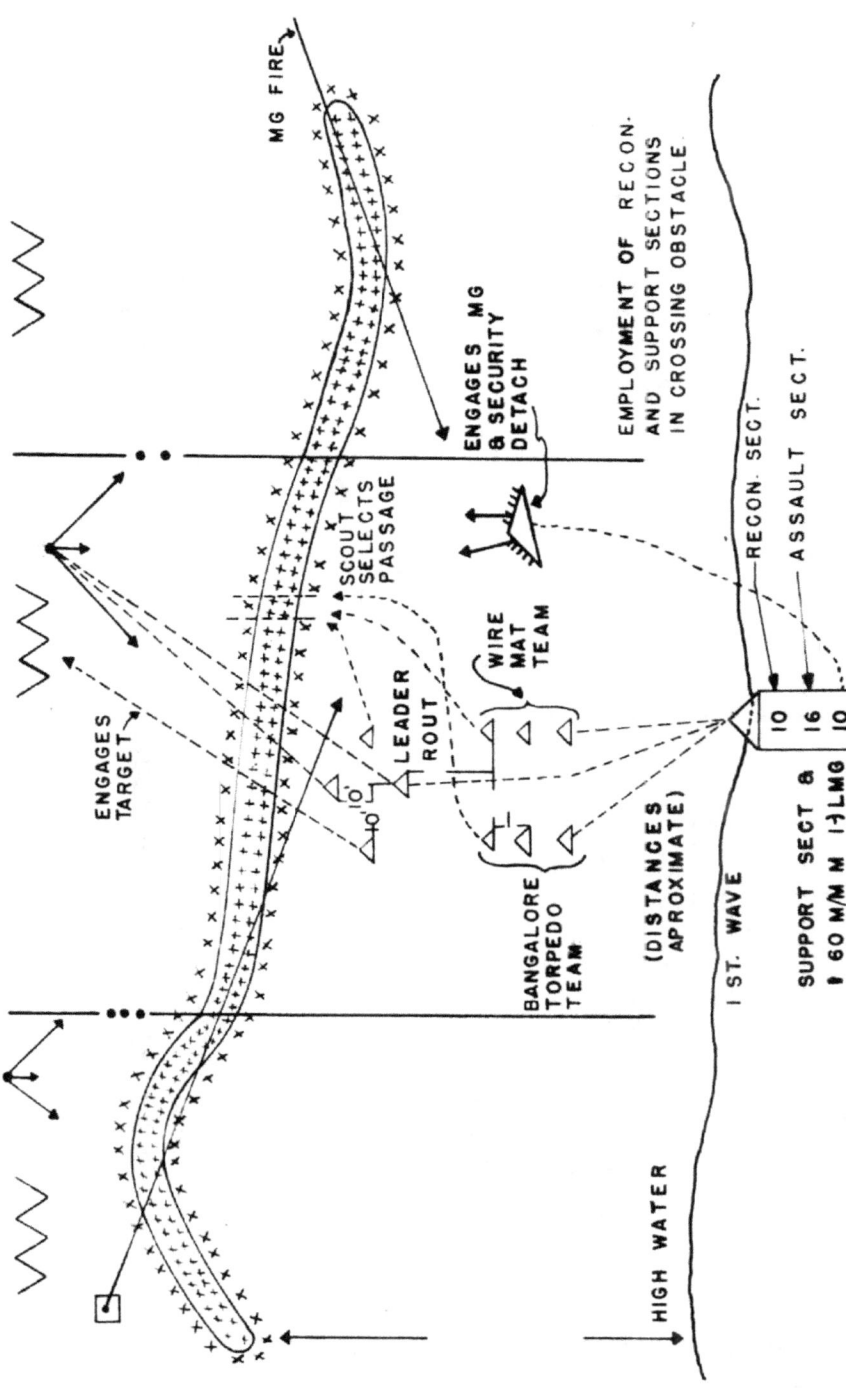

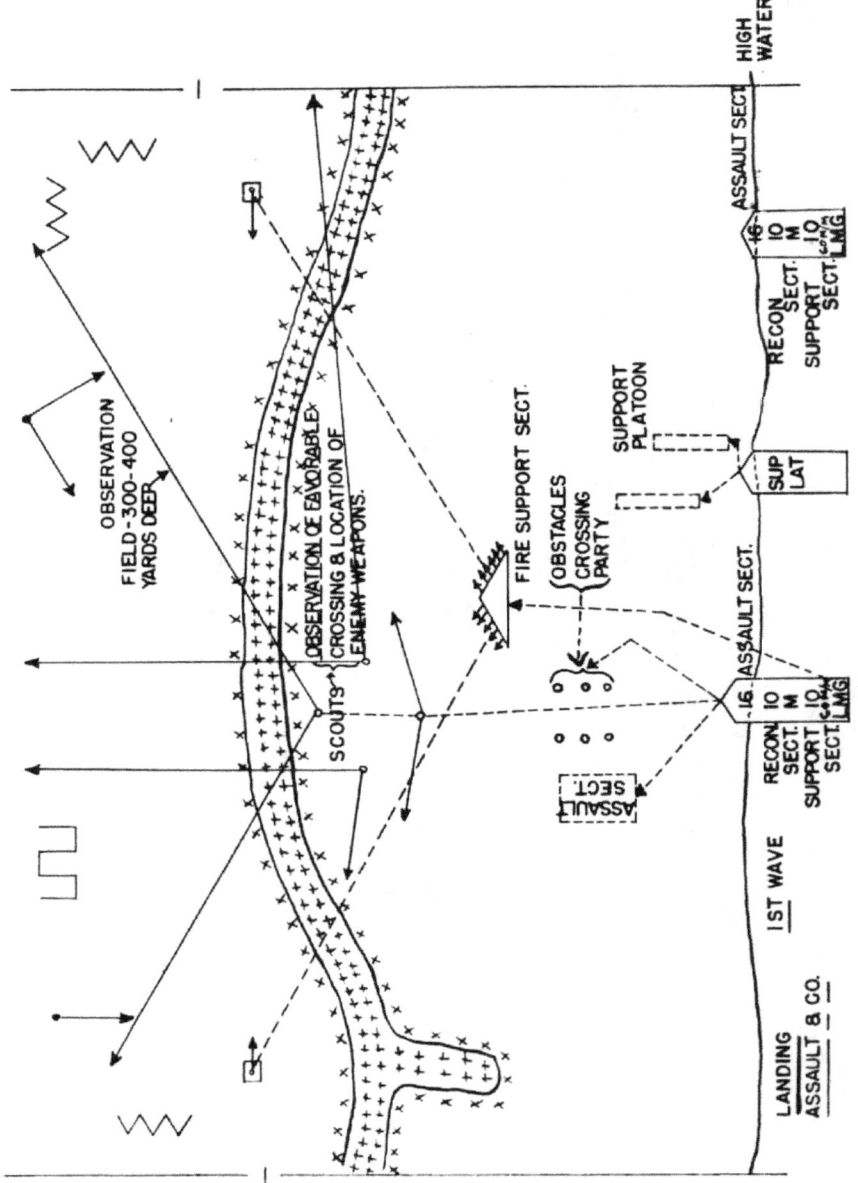

(7) Machine guns in fortified positions usually include weapons sited in open emplacements, having missions of protecting pillboxes, wire obstacles, minefields, etc. The relative ease with which these machine guns can be concealed makes them quite difficult to locate, especially during heavy firing of artillery and air attacks by both the enemy and friendly forces. As the fire of only one well concealed enemy machine gun may seriously delay the entire operation, Assault Platoons should be thoroughly trained in promptly locating and reporting such weapons. Machine guns in open emplacements will in general be engaged as follows:

As the Scout Party advances toward the enemy position, or after having halted temporarily in front of an obstacle to observe it, it may discover the location of a machine gun firing from an open emplacement. (While searching for the machine gun, the party should also seek to locate the position of the local security detachments protecting the hostile machine gun). If it appears that the fires of the machine gun and the members of the hostile security detachment may prevent the establishment of a passage thru the obstacle, the scouts will first attempt to knock out the machine gun crew. If it is not possible for them to engage the weapon, they may signal to the Support Section and indicate the target by the use of tracers, advising whether the target is to be engaged at once or on signal later, in accordance with plans. Fires of the Support Platoon, especially mortar fire, and the first of the accompanying tanks will also be utilized.

If the hostile fires will not interfere with the establishment of a passage, or if it is not practical to engage the machine gun from the positions available to the Support Section, the Scout Party will pick out some point within the enemy defense area from which effective fire can be directed on the emplacement when friendly troops are in a position to do so.

The Assault Platoon will not normally be used against machine guns sited in open emplacements.

(8) The reduction of pillboxes is the special mission of the Assault Section, which is the only unit of the Assault Platoon equipped with adequate means to accomplish the task. Assistance in the reduction of the pillbox is provided by all sections of the Assault Platoon, particularly the Support Section, whose primary mission is to engage hostile weapons which interfere with the operations of the Assault Section. <u>The Assault platoon will be closely supported by direct gunfire of accompanying tanks and various types of support craft.</u> These fires must be effectively coordinated.

The location of hostile pillboxes, particularly in the initial stage of landing, will be done by the Scout Party, which will pass the information on to the Assault Section. Upon receipt of information giving definite location of hostile pillboxes, rapid revision or modification of existing plans may be made by the Assault Platoon Leader,.

When the protective fires of hostile pillboxes are knocked out, or neutralized by friendly fire support, the Assault Section moves under cover of the fires to positions from which the reduction of the pillbox is undertaken in accordance with prepared plans.

(9) Thus, the Assault platoons in the first wave accomplish such silencing of emplacements as is required to pierce the initial defenses and sufficiently interrupt the continuity of defensive fire bands, to permit succeeding assault troops to cross the beach without prohibitive casualties.

d.) <u>Second Wave</u> - Assault troops in the second wave silence additional emplacements. If the scheme of maneuver is a frontal assault of the desired landing beach, the second and following waves widen the initial gap sufficiently to permit exploiting troops to get inland from the beaches, without fighting their way through or past beachline defenses. If the scheme of maneuver is the seizure of a lightly defended beach, from which to envelop the desired beach, part, or possibly all, of the second wave roll back the inside flank to permit enveloping troops to get through the immediate front, without stopping to fight and without receiving undue casualties from direct fire. The balance of the second wave may consist of the leading echelon of enveloping troops; or it may consist of platoons of the reserve company, which proceed to positions from which to protect the outside flank of the envelopment. Both missions are urgent and their comparative urgency can be determined only by analysis of the elements of a specific situation. In any event, the battalion commander must be alert and bold in the use of his support company to seize and defend the perimeter of a small battalion beach head. A toe-hold on a beach is precarious. Every foot of terrain with defensive advantage that can be seized inside the beach lessens this precariousness.

e.) <u>Battalion Base of Fire</u> - The battalion commander must build up his battalion base of fire ashore as early as possible, in order to deal with hostile mortars and other heavy weapons. In addition to the weapons of the Heavy Weapons Company, the 4.2 inch chemical mortar, firing HE shell, is believed well adapted to this purpose. Self propelled artillery may be attached to the battalion, and should be landed as early as possible.

f.) <u>The Regiment</u> - The regimental combat team commander utilizes his Reserve Battalion to strike rapidly inland, over-run hostile mortar positions and seize good defensive terrain on the regimental intermediate beach head line in his zone of action. He must build up his base of fire for support of this action and support of subsequent defense against counterattack, by early employment ashore of his attached artillery.

g.) <u>The Division</u> - The advance of the dominant terrain on the beach head perimeter is made by the Division Reserve Combat Team. Parachute troops, landing just before daylight and seizing dominant terrain on the division objective, may greatly facilitate the action of the division reserve.

h.) <u>Maneuver</u> - Throughout the action reserves are used to exploit success and advance the action inland. Commanders of all echelons must remember the race between the build-up and the movement of hostile main reserves. The beach head is essential to the build-up. Troops holding the beach head must have time to get braced before they are hit by a coordinated attack and they must have fire support in position, ready to fire. Only that energy is expended on the beach defenses which will facilitate inland advancement of the action and its support. Isolated defensive installations, not seriously interfering with operations on the used beaches, are dealt with later. Inland advancement is accomplished by reserves, fresh, organized and under control. Initial assault units will be somewhat exhausted and disorganized. They assist the advance of reserve units by fire on the flanks of the gap and may be reorganized to mop-up isolated sections of the beach defenses. Inland advancement with necessary speed requires alertness to opportunity, and boldness in the commitment of reserves.

i.) The problem of reducing the beach line fortifications and proceeding to the establishment of a beach head is the total problem within the scope of this circular. Reference is made to FM 31-5 for treatment of amphibious operations, to FM 100-5 for applicable doctrines of the combined arms, and to T.C. No.33 for doctrine regarding the attack of a Fortified Position. Familiarity with the provisions of these manuals and the field manuals of the arms and services concerned is essential.

The example that follows demonstrates the foregoing tactics and technique. Tactics, technique and example are presented as a guide. They must be moulded to fit the specific situation. Deviation from these guide lines is a command prerogative which the able commander will exercise and justify by the results obtained.

As stated in Sec V, the composition of an Assault Company may be varied to fit a particular situation. In this example a Company with four assault platoons has been used.

4.) <u>Example</u>: a.) The 10th Division (reinforced), assault division of the XVI Corps, has the mission of attacking a strongly fortified beach between the corps boundaries (as shown in Fig 1), seizing, defending and maintaining the beach head in the corps zone of action.

    (1) Attached to the 10th Division:

        (a) Tank Bn. (M)
        (b) Tank Destroyer Bn
        (c) CML Bn (Mtz)
        (d) 2 Antiaircraft Bns., AW
        (e) 2 Ranger Bns
        (f) Shore Party Group.

b.) Supporting operations under control of higher echelons consist of:

    (1) Air bombardment and naval gunfire support
    (2) Seizure of dominant terrain on division intermediate beach head line by parachute troops.
    (3) Delaying action against advance of hostile main reserves by glider-borne troops.

c.) Analysis of the situation confronting the 10th Division: Examination of terrain shows Able Beach to be about 3,500 yards long and Baker Beach about 600 yards. There is a good exit from Baker Beach, and one from Able Beach at Able Beach Red. Landing craft can beach at any point on either beach. Remainder of coast line in corps zone of action is rocky with steep cliffs, negotiable only by Rangers landing from rubber boats.

Known defenses are as indicated in Fig 1. It will be noted that Baker Beach is strongly defended throughout, and that Able Beach is strongly defended at its extremities. Defenses in the center of Able Beach are comparatively weak.

d.) <u>Scheme of Maneuver</u> -

(1) The division commander decides that he must have both exits in order to accomplish his mission. He accordingly decides to employ two combat teams abreast, one to seize the beach served by each exit. Baker Beach is flanked by rocky headlands and must therefore be assaulted frontally. On Able Beach it appears advantageous to penetrate the weaker center defenses initially and destroy the stronger defenses at the exit from the flank and rear, utilizing foot troops with hand carried weapons. This scheme of maneuver enables the attacker to maintain neutralization of defense areas at the extremities of Able Beach by naval gunfire, air bombardment, and smoke until a foothold has been gained and he is prepared to assault these defense areas. The defense area at the south end of Able Beach will be assaulted from the rear by CT28 after it has gained Baker Beach.

Ranger Battalions land under cover of darkness, approach by stealth and destroy the shore batteries on the two headlands flanking Able Beach. One regiment of parachute troops will seize dominant terrain on the nose of the division intermediate beach head line.

The chronological sequence of the different phases of this attack might be as follows:

H-hour is one hour after nautical twilight and two hours before high tide.

H minus two hours: Rangers land to destroy batteries on beach lands.

H minus two hours: Parachute troops drop.

H minus 60: Naval bombardment on shore batteries starts, only if batteries open fire.

H minus 30: Air bombardment starts.

H minus 20: Planes start laying smoke screen.

H minus 15: Parachute troops attack and glider elements of airborne division begin to arrive.

H minus 10: 4.2 inch mortars on landing craft assist in laying and maintaining smoke screen.

H-Hour: Leading wave lands.

The Division Commander's scheme of maneuver is shown in Figure 1 (operations overlay)

(2) Combat Team 29 capitalizes on the available length of weakly defended beach by landing two battalion teams abreast, thus providing for prompt inland advancement of the action by its right battalion. The left assault battalion lands on Able Beach Green, disposes of defenses in its front and proceeds to the attack of the defenses of Able Beach Red from the flank and rear. The lifting of supporting fires on these defenses is coordinated with the attack by this battalion.

The Reserve Battalion, prepared to land on either Able Beach Yellow or Red, lands and advances rapidly inland to the regimental intermediate beach head line on the left of the zone of action of Combat Team 29, (see Figure 1).

Combat Team 28 is compelled by the length of beach and strength of defenses to land with battalion teams in column. The leading battalion will penetrate the defenses and obstacles on Baker Beach and advance to the initial phase line, prepared to resist local counter-attack and protect the flank of the next

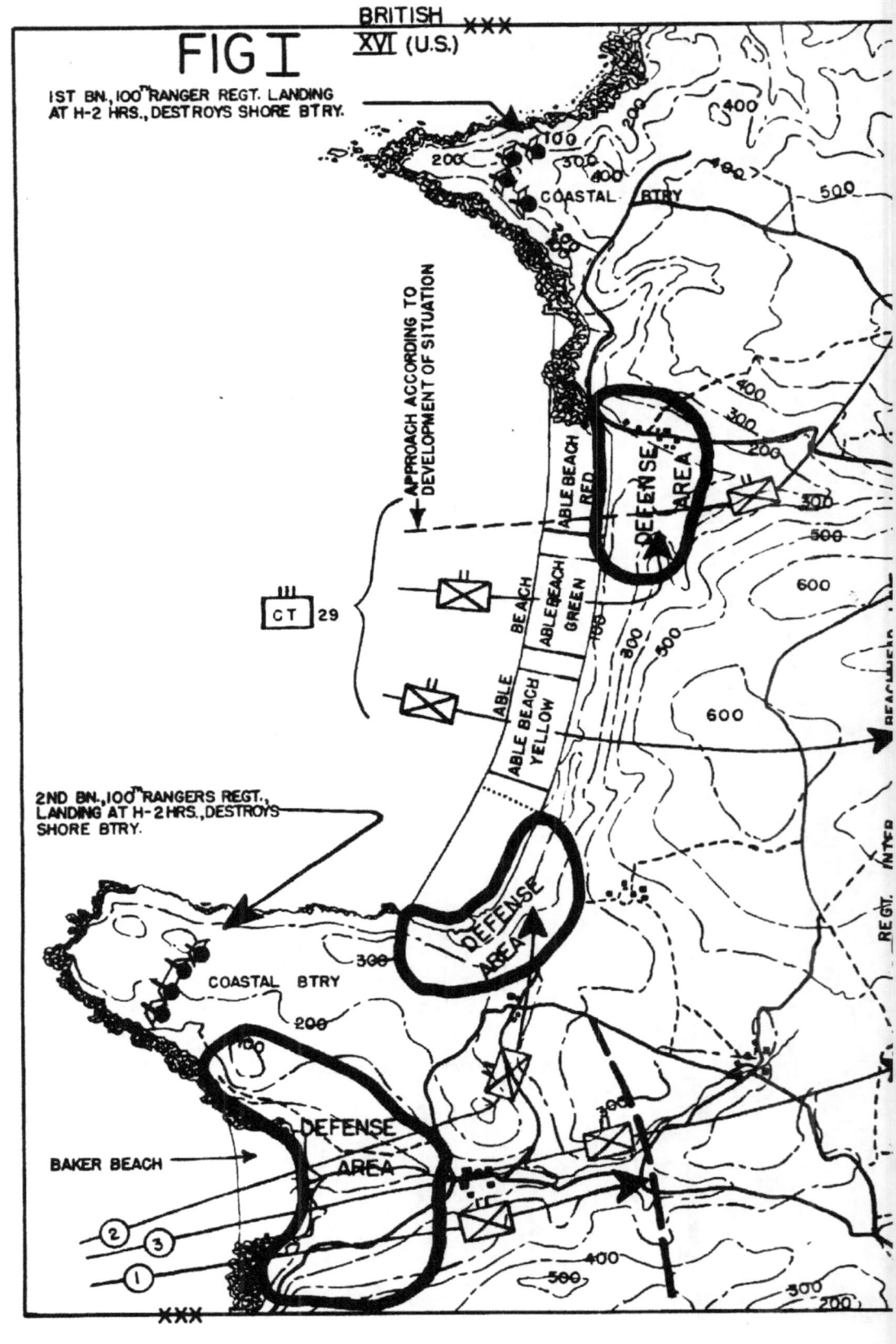

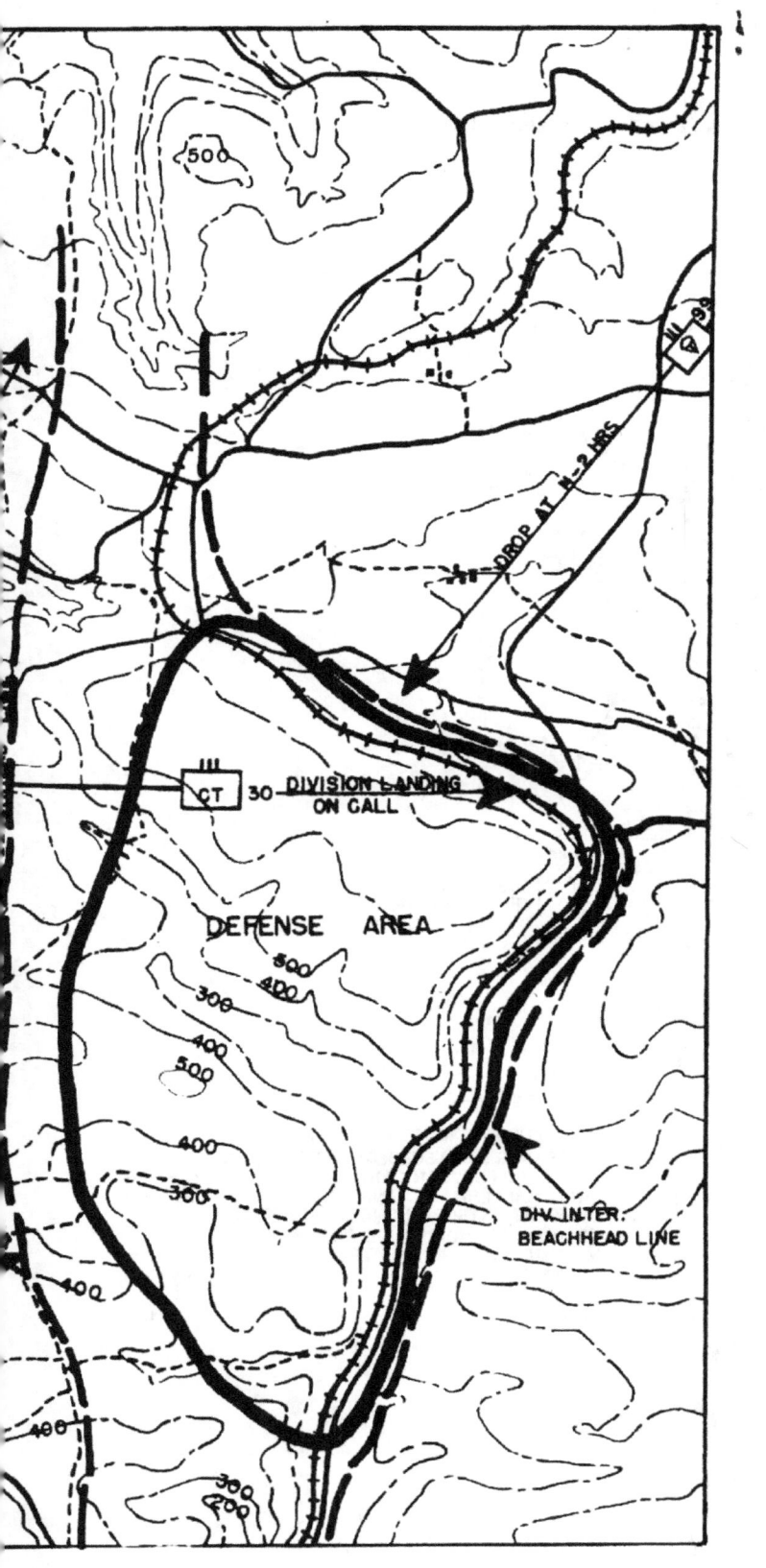

FIG 2

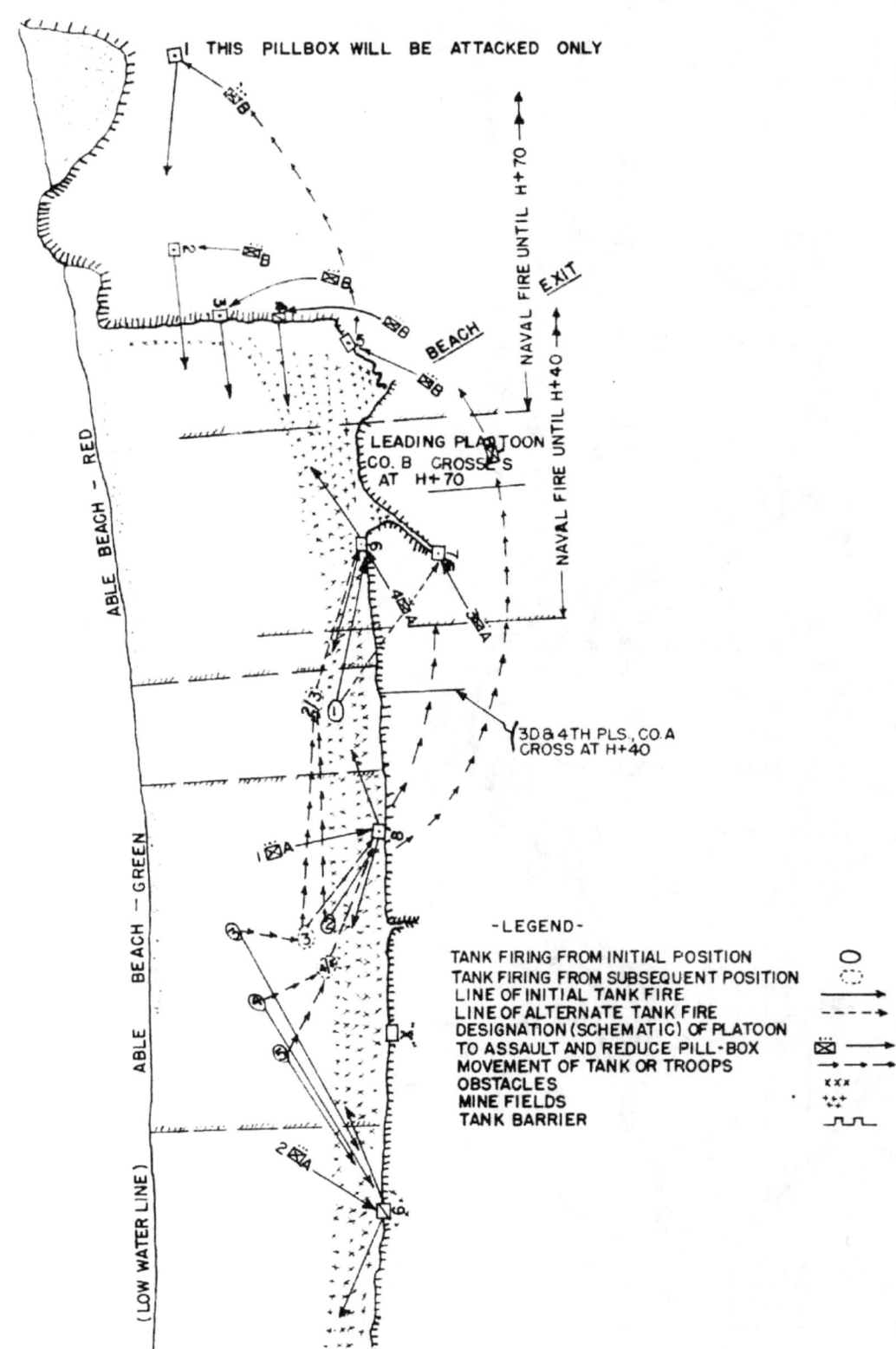

battalion during its attack on the defense area at the south end of Able Beach. (Figure 1). The reserve battalion lands and advances rapidly inland to seize and hold the regimental intermediate beach head line in the regimental combat team zone of action.

e.) <u>Details of Assault</u>: The assault of the leading elements, left assault battalion of CT 29 will be examined in detail. (See Figure 2).

Neutralization of pillboxes 1 to 7, inclusive, will be maintained by naval gunfire during the initial phase of the assault. Pillboxes south of No.9 will be accounted for by the right assault battalion. Pillbox marked "X" represents a possible additional box, discovered only after landing and located not necessarily as shown, but anywhere along the beach defense line.

The battalion consists of two assault companies and one reserve company. The assault companies each contain four assault platoons. A company of tanks is attached to the battalion.

The battalion commander decides to attack in a column of companies. The first wave consists of two assault platoons supported by a platoon of five tanks. Under protection of direct fire on embrasures, delivered by tanks, the 1st Platoon, Co "A" assaults pillbox No.8 and the 2nd Platoon, pillbox No.9.

The second wave consists of the balance of Co "A" (Co Hqs and 2 platoons) and Co "B", less 2 platoons. The 3rd and 4th Platoons of Co "A" cross the beach in gaps created by the first wave, under protection of tank fires, and proceed to the assault of pillboxes 6 and 7. Naval gunfire and air bombardment shifts to include only pillboxes 1 to 5 inclusive.

Co "B" less 2 platoons, crosses the beach at the same time, and, swinging inside the 3rd and 4th platoons of Co "A", begins its approach to the attack of the defense area point covering Able Beach Green. The remaining 2 platoons of Co "B" assault the Able Beach Green defenses, each platoon closing with a specifically designated pillbox. Naval gunfire lifts when Co "B" approaches. Tanks for the support of this attack land in a later wave.

The order and timing of boat waves is derived from time and space estimation of the operation ashore. Troops in the second wave should pass through the gaps in the defensive fire band as soon as the gap is created by the reduction of pillboxes 8 and 9; not before. Fifteen minutes is allotted for the reduction of pillboxes 8 and 9. If the second wave lands at H plus 12, it is believed the gap will be created by the time it reaches the zone of defensive fire bands. Landing at this time, the points of the 3rd and 4th platoons, Co "A", should be within about 400 yards of pillboxes 6 and 7, by H plus 40, at which time, therefore, the schedule of naval gunfire provided that fire would shift to include only pillboxes 1 to 5, inclusive. Co "A" should be within 400 yards of an east-west line through pillbox 5 at H plus 70; therefore naval gunfire is scheduled to lift at this time. Tanks in support of Co "A" must be landed and ready to fire from the beach on pillboxes 2, 3, 4 and 5, when naval gunfire lifts; therefore these tanks are landed at H plus 1 hour. The third wave should land in time to enable Co "B" to proceed to its mission with no delay brought about by awaiting its 3rd and 4th platoons. Its landing time is set at H plus 18. Remaining boat space in the third and later waves is utilized in landing the reserve company, battalion base of fire, and necessary elements of the Shore Party.

In event of receipt of fire from a previously undisclosed pillbox, necessary change in plans must be made on the spot to provide for its destruction. If pillbox "X" (figure 3) develops, fire of an additional tank is placed on it as soon as box No 8 is silenced. The 4th platoon, Co "A" might be directed to the assault of this box, in which event one platoon of Co "B" may be diverted to No.7. The deficiency in Co "B" may be made up by reinforcement from reorganized elements of Co "A". Prepared plans should include provisions for likely contingencies.

f.) <u>A prearranged and rehearsed plan of tank fires</u> provides for engagement of targets with maximum speed and minimum confusion. Reference is made to assignment of targets shown in Table 1 for an example. The tabulated assignment was arrived at by means of the following analysis. Initially, three tanks are placed on the antitank pillbox to insure prompt engagement by one. Then, assuming the ideal of no loss of tanks, the most conveniently landed tank is placed on a machine gun pillbox to cover each pillbox by the fire of one tank. The remaining two tanks are kept on the A-T pillbox. This sets up the second targets. Now it is assumed that two of the five tanks are destroyed. Pillboxes 6, 8, and 9 are most dangerous and must therefore be kept covered. Table 2 shows assignment of tanks to these targets, on the basis of loss of any two tanks. Assignments of targets to tanks is extracted from Table 2 and compiled in Table 3. Targets already assigned are removed from the assignment and the net additional targets shown in the last column. These are set down in Table 1 as alternate targets. We may now obtain a bonus by discovering upon landing that naval gunfire has destroyed one of the more dangerous targets. Target 8 is assigned as an alternate to Tank 5 to cover this contingency and target 7 to tank 6. Finally, a previously undisclosed target "X" may be discovered anywhere. Such target is therefore assigned as an alternate to the tanks with one other alternate target.

The assignment of targets to tanks 6 to 10 inclusive (2nd platoon) is worked out in exactly the same manner.

TABLE I

Tank No	Initial No	Target Type	Second Target No	Type	Alternate Target No	Type
1	6	MG	6	MG	7/X	MG/AT
2	8	MG	8	MG	6/X	MG/AT
3	9	AT	7	MG	6/8	MG
4	9	AT	9	AT	8/X	MG/AT
5	9	AT	9	AT	8/X	MG/AT
6	2	MG	2	MG	1/X	MG/AT
7	3	MG	3	MG	2/X	MG/AT
8	4	AT	5	MG	2/3	MG
9	4	AT	4	AT	3/X	MG/AT
10	4	AT	4	AT	5/X	MG/AT

TABLE II

Target	Tank
6	1 1 1 2 1 1 2 1 2 3
8	2 2 3 3 2 3 3 4 4 4
9	3 4 4 4 5 5 5 5 5 5

TABLE III

Tank	Target	Previously assigned	Additional
1	6	6	
2	6-8	8	6
3	6-8-9	9	6-8
4	8-9	9	8
5	9	9	

 g.) The plan presented in paragraphs e.) and f.) above, is intricate and in detail. Attack of a line of concrete fortifications involves employment of specific units and sources of fire on specific emplacements. Confusion can be avoided only by specifically detailed plans. He who hesitates and ponders on the beach promptly ceases to exist.

 Execution of such plans can be successful only if rehearsed over and over against full scale reproductions of all that is known of the defenses on the beach to be assaulted. On the other hand, commanders must avoid becoming so frozen to fixed lines that they are unable to make prompt changes to meet the contingencies of battle. All echelons must be kept alert and flexible through introduction of such contingencies in rehearsals. All echelons must learn to anticipate various contingencies and have a plan in mind with which to meet them.

 h.) The foregoing is a solution, not necessarily the last word. It demonstrates the need for specific planning and time and space estimation. It is not to be copied, but rather to be used as a starting point and a guide in planning and training for assault of a fortified beach line. Each commander must analyze his specific situation and plan accordingly. He must THINK!

 5.) <u>Time of Attack</u>: While the attack illustrated took place after nautical twilight, circumstances in a particular situation may indicate that the attack be made at night.

 Sec V, Chapter 1, FM 31-5 presents the considerations affecting the decision as to time of attack. Some minor modifications in assault technique may be required for night attack.

VII. SPECIAL TRAINING.

 The complexities of assault landing require special organizations and equipment and these, in turn, require specialized training. The details of this training are discussed adequately in FM 31-5. One cannot over emphasize, however, the importance of full-scale rehearsals as the culminating phase of training. Higher commanders, in their planning, must recognize the importance of rehearsals and take positive action to insure that sufficient time and facilities are allotted for their execution.

VIII: APPENDIX

A.) USE OF SMOKE IN THE ASSAULT OF A STRONGLY DEFENDED BEACH

1.) **Types of smoke weapons and munitions**: Smoke may be put down by the following means:

    a.) From aircraft by means of bombs or smoke tanks;
    b.) By shellfire from mortars or field guns;
    c.) By smoke generators, either fixed, mobile or floating;
    d.) By smoke grenades.

2.) **Aircraft**

    a.) There are two general types of smoke bombs that can be used from aircraft - the continuous burning type (HC or HC Substitute) and the instantaneous and non-continuous type filled with WP. A WP-filled bomb has the advantage of also being a casualty agent due to the incendiary effect of the burning particles striking defenders, but has the disadvantage that the time of emission of the smoke is not predictable. The HC smoke bombs can be counted on to produce a screening smoke for the predictable time of 5, 10 or 20 minutes.

    b.) Smoke tanks (British S.C.I - Smoke curtain installation): Fighter aircraft and light or medium bombers can carry either wing or bomb bay smoke tanks. The liquid FM or FS (British CSAM) is discharged from the tanks while flying over the area to be screened. The length of the screen depends on the speed of the aircraft and the capacity of the tank.

    c.) Float smoke bomb: this munition is under development. The bomb will function either from the land or water. It is expected that it will be in supply in the near future.

3.) **Shellfire**:

    a.) Mortars are used for the close support of infantry units, and can establish and maintain a screen as long as the munition supply is ample. The two weapons for which smoke ammunition is available are:

        (1) 4.2" Chemical Mortar, MIAI
        (2) 81 mm Mortar, MI.

The number of rounds required is dependent on the width of the target and wind direction at the target, and also on whether a casualty effect is desired. The smoke (WP) capacity of the 4.2" mortar shell is 7.56 lb and of the 81 mm mortar shell, 4.06 lb. The 4.2" chemical mortar is fired by chemical troops, and can also fire HE: the 81 mm Mortar is fired by the infantry.

    b.) Artillery: In the assault of a beach, artillery usually will not be available in the initial stages. Tanks can fire a smoke shell. British tanks are equipped with a breech loading mortar which fires a smoke shell. The utilization of the tanks will be best effected by their employment in their proper tactical role and should not be counted on to provide a smoke screen.

4. **Smoke Generators**:

    a.) Smoke pots containing HC or CTC (Carbon tetrachloride

composition - British) are available in many sizes and types. The burning times of these munitions vary with the weight and type of filling. The pots could be man-handled to a flank and fired to screen movements. A floating type is available which can be launched from assault craft. The screen can be maintained by firing additional generators as the original ones are burnt out.

b.) Mobile generators are mounted on trailers and provide a screen by the vaporization of a fog oil in a furnace. These generators are designed for long-time operation and produce a dense screen which holds together and moves with the wind just above the ground. A type known as the 'Esso' generator is available and could be operated from landing craft to provide cover from air attack for a beach operation. The use of these generators would be best made in covering subsequent landings after the initial assault has cleared the beaches.

5.) Smoke Grenades: These weapons produce a large volume of smoke for two minutes and can be thrown by the first wave onto the beach. A rifle grenade discharger is under development for this grenade.

6.) Tactical use of smoke:

a.) Time: Smoke screens in general require a time to establish and must be maintained. All operations which are dependent on smoke for their success must allow for the establishment and maintenance of the screen before being put into effect.

b.) Obscuring power: The value of smoke is in the ratio of 40:12:3, that is 40% hits without smoke, 12% when target only is obscured and 3% when the weapons are obscured. With this in mind, it must be realized that drifting screens (curtains), after initially covering the enemy, may move with the wind and obscure the attackers if the attack is not timed to coincide with the smoke effect. The effect of smoke is valuable on moonlight nights as well as during daylight, but the expenditure of munitions during night is much less. The maximum use of obscuration is made when the smoke blinds both the aimed small arms fire of the defenders as well as the enemy observation for artillery fire.

c.) Signals: Walky-talky radio communication should be available to all units and air-ground communication should be immediately available to ground force commanders as soon as they are in a position to direct the assault.

d.) Use of Aircraft:

(1) The initial screens will be placed by aircraft. In order to insure a satisfactory screen, it is of primary importance that the smoke-laying aircraft be given protection from small arms fire. This must be done by smoke bombs of a continuous burning type being dropped to provide a major degree of protection from aimed small arms fire so that it cannot be brought to bear successfully on all aircraft which are committed to a smoke mission. Time must be allowed for the protective screen to form before the curtain type of screen is laid over the beach defenses. Dependent on the wind, it may be necessary to re-establish the initial screen over the defenders before additional curtains are laid.

(2) Aircraft laying smoke curtains must be afforded fighter cover, and spare smoke laying aircraft should rendezvous so that they will be available for prompt usage should the situation warrant their employment.

(3) Inasmuch as the use of smoke is of primary importance to the ground forces, the aircraft should be under call of the air support party, who will utilize the aircraft as the situation develops, for the ground commander.

(4) The number of aircraft to be employed on a screening mission should be determined by the air officer in conjunction with the air chemical officer for the initial assault. Reserve aircraft (smoke) should be thereafter under the control of the ground force commander until released.

e.) Use of Mortars:

(1) Mortars are the chief means of producing smoke after the initial landings have been made. A high priority in landing should be afforded mortars and mortar ammunition in order that maximum use may be made of this weapon. The supply of ammunition will present the most difficult problem in the first stages of the assault.

(2) Smoke from mortars should be used to cover the approach of a unit when the advance is over open ground. Exposed flanks of units should be covered whenever threatened by enfilade fire. As the supply of ammunition will be limited, smoke should be used with economy and only during the critical stages of an assault.

f.) Use of smoke pots and mobile generators: the use of these munitions and equipment is governed by their availability and the necessity for their use in screening large areas. Their employment should be under the control of a chemical officer and should be coordinated with the antiaircraft officer. Their utilization is limited by the time required to establish a satisfactory screen and therefore can only be employed when an aircraft warning system is available. It is not possible that these munitions could be efficiently and satisfactorily used until 24 to 48 hours after the initial assault has cleared a port or beach.

g.) Depending on the wind direction, if onshore, floating smoke pots can be used to screen the flanks of assault battalions. The smoke floats can be transported in assault craft and dropped overboard, where they will burn for 12 minutes. Allowing the time necessary to establish the screen, the flanks can both be protected if an onshore wind is blowing.

h.) Support from boats: Low trajectory weapons of the supporting craft, both naval and landing type will be able to furnish some smoke cover in the initial stages of the assault. However, in the latter stage as the assault craft approach the beach their fire will have to lift and dependence made on aircraft smoke screens. Fire from mortars on assault boats during rough water will not be dependable and may well be a disadvantage.

B.) Use of Flame Throwers in the assault of a strongly Defended Beach.

1.

a.) At present the portable Flame Thrower M1A1 is

the only weapon standardized and in supply to the American Forces. This is a pack type model with a capacity of four gallons of fuel and weighs about 70 lbs. filled. It is operated by one man and has a maximum effective range of 50 to 70 yards and fire continuously for 12 seconds or in small bursts. The small bursts are the most effective as they enable the maximum range to be obtained with the greatest effect as change in aim is possible between shots.

    b.) The British one hundred (100) Crocodile type flame throwers may be mounted on Sherman tanks. The fuel for this flame thrower, together with the nitrogen gas for the pressure is carried in an armored trailer connected to the tank by a universal joint. Four hundred gallons of fuel, sufficient for 20-three second discharges are carried. The trailer can be jettisoned by the crew without leaving the tank when the fuel has been expended. The tank maintains all its regular armament and after the trailer has been jettisoned it can perform its normal function. The mobility of the tank is only slightly restricted by the trailer. The range of the flame thrower is:

        Maximum   200 yards
        Maximum effective   140 to 160 yards.

2. Tactical use of Flame Throwers:

    a.) The development of thickened fuel enables the flame thrower to be used at greater ranges than is possible with ordinary liquid fuels. The fuel is not consumed in flight and burns after it strikes the target. The thickened fuel is difficult to put out and burns without smoke. The morale effect of large flame is lost as well as the lack of smoke; these disadvantages are more than overcome by the range and continuous burning effect on the target.

    b.) The pack type flame thrower is utilized for the reduction of field fortifications to provide close-in protection for the placement of demolition charges. The flame throwers must always be supported by other weapons, fire or smoke. Operators should be thoroughly trained as well as replacements or reserve operators, otherwise operations centered about its use will fail. A two-man team should normally be assigned each weapon.

    c.) Each weapons team is normally assigned to an assault detachment. The assault detachment moves forward under supporting flat trajectory fire, and approaches a pill box on its blind side. Supporting fire from adjacent pill boxes must be neutralized before the assault detachment can approach the pill box. When the assault detachment is in position, the supporting fire is signalled to lift and the flame throwers rush the target, and blind and burn the gun crews. The demolition men then place their charges and all take cover for the detonation. After mopping up the detachment reorganizes.

    d.) All equipment must be carefully checked before use; only from experience can the operator know the amount of fuel remaining after a partial discharge.

3.) The principle operations of the tank-type flame thrower presupposes that the bulk of the enemy organization has been broken and the flame thrower would be used for mopping up, trenches, bomb holes, pill boxes or houses. The tank can deal with 100 feet of trench with one or two shots lasting 2 to 4 seconds. The filling effect of a quantity entering a loophole will incapacitate the occupants of a pillbox. Likewise, machine gun posts near beach landings can be quickly dealt with from the flanks.

4.) The principle usage of the tank flame thrower is therefore as an accompanying tank. As soon as the beach minefield has been cleared, flame thrower tanks should be landed to assist the infantry in rolling back enemy flanks in order to extend the beach and landing areas.

C.) <u>Chemical Warfare during the assault of a strongly Defended Beach.</u>

1.) It should be expected that during and after the assault of the beaches chemical warfare will be begun. Chemical warfare combines flame, smoke and gas.

2.) The principle beaches may in all probability be protected with emplaced flame throwers. Destruction of this equipment by bombing and gunfire must be undertaken in order to make a successful landing.

3.) The approaches may be mined with vesicant mines H5 (mustard) or possibly large emplacements of CG (phosgene) which can be released at the will of the defenders. Destructions of these installations will have to be accomplished by bombardment or shellfire.

4.) Passage of gassed and contaminated areas will have to be made, relying on the protection afforded by the mask, protective clothing, shoes (properly and freshly impregnated) and protective ointment for personal decontamination. The British Cape, Anti-gas is to be preferred to the AS cover, protective individual, for protection against spray of vesicants from the air, as the British cape allows normal movement of the individual and affords the necessary protection.

5.) The concentration of troops in a landing area is a desirable target from the enemy's point of view for a spray attack. It should be expected that desperate attempts will be made to spray attack our troops. Although our equipment is satisfactory, there is the probability that large numbers will become casualties after a spray attack and this will be all the more to our disadvantage with the lack of decontamination equipment for clothing, lack of replacement clothing and lack of shelter until the beach head is established. Gas training, gas discipline and personal decontamination should have a high priority in all training programs.

6.) All ammunition and food brought ashore in the first stages should be in sealed containers to prevent contamination. It will be increasingly difficult to bring ashore, distribute and utilize the quantities of bleach necessary for decontamination. Dispersal of equipment and temporary abandonment of all except high priority items will be necessary until the chemical

Photograph # 1

Fixed Flame-Throwers emplaced on beach

Photograph # 2

Fixed Flame-Throwers emplaced in the water.

Photograph # 3

General effect of use of Flame Throwers on beaches.

Photograph # 4

Close view of effect of Flame-Throwers on beach.

decontamination companies can be landed. All service units in the initial assault should be trained to decontaminate equipment using simple or improvised materials.

7.) Use of gas by our own troops would ordinarily not be a suitable employment of our ground forces. The Air Chemical Officer should advise the operations commander and Air Force Commander on the employment of the air arm for relation and casualty effect. The necessity for our troops passing through areas which we ourselves may have contaminated is a basic consideration that must be taken into account on every decision to use vesicants.

BEACH OBSTACLES

Photograph No. 1

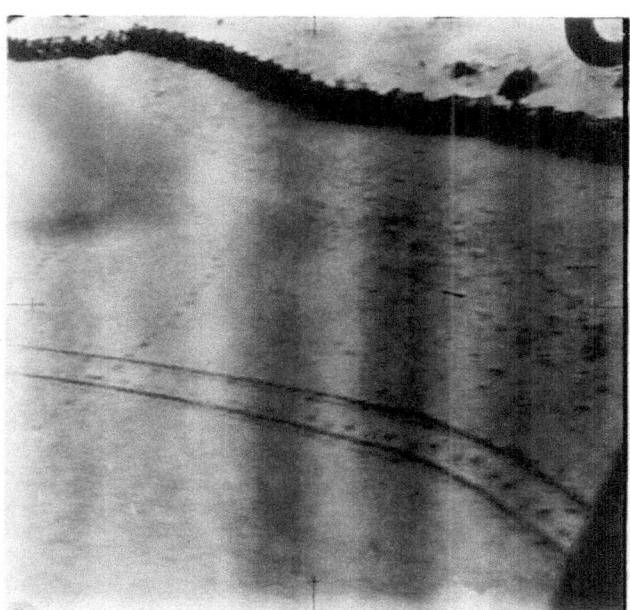

Photograph No. 2

Beach obstacles, wire: continuous belts of wire along all open beaches, usually sited between high water mark and the back of the beach, as above.

BEACH OBSTACLES

Photograph No. 3: Wire atop of Promenade.

Photograph No. 4: Reinforced concrete walls

BEACH OBSTACLES

Photograph No. 5: Steel fence obstacle, along open beach

Photograph No. 6: Walled up houses on sea-front of a town.

CONCLUDING REMARKS

ASSAULT TRAINING CENTER

CONFERENCE

CONCLUDING SESSION

HQ. ETOUSA.

23 June 1943

SUMMATION BY: THE CHAIRMAN, LT. COL. L. P. CHASE.

This is the last session of the Conference. Let us consider the record of the proceedings.

We have here copies of all the talks which have been edited to date. There are a few insert sheets which indicate speeches, not yet corrected and returned. They will be included as soon as they have been edited. We have also the conclusions reached as the result of each day's discussion, as approved by the Conference from day to day. We have incorporated also summaries of some of the discussions which did not reach the status of conclusions. Additional records of discussions are now under preparation.

These papers were issued to you in this draft form, so that you could run through them quickly, with especial attention to the parts you are interested in. This concluding session is your opportunity to raise any questions in connection with the proceedings. When the proceedings are put out in final form, it will include several things not now in this draft edition: foreword, table of contents, marginal index, and appendices; to make them more readily usable. We do not expect anybody to sit down and read the proceedings from cover to cover. It is intended to be simply a reference work for those interested in a particular phase of the Assault Landing.

At the conclusion of today's discussion, we would like to have all of the tentative proceedings returned. The final edition will be ready for mailing about a week from now. Every member of the Conference will receive one copy and all interested agencies, both American and British, will be given additional copies. We are preparing a distribution list and should be glad to have your suggestions.

The training circular which Col. Lock's Committee has been preparing will be included as an appendix in this Record, in addition to being distributed separately as a tentative training circular. Comments on FM 31-5 will likewise be included in the appendix.

This is the occasion for a formal approval of the Proceedings details which have all been approved previously in the Conference. If there are no objections, the proceedings will stand approved, subject to the additions and corrections which have been mentioned here.

CONCLUDING REMARKS BY: BRIG. GEN. D. NOCE, A.C. of S., G-3, ETOUSA

A year ago when I got into this work actively, there was

very little that could be found dealing with amphibious operations. It was found, in the U.S., that the best we had was that of the Marine Corps. We have gone a long way since then. The officers who spoke before this Conference were the best informed people on the lines on which they spoke that I have yet seen assembled. We had Hughes-Hallet, the Commodore who commanded the Naval Force at Dieppe, and I think, was at Dunkirk. He knows this business from his own experience. We had Major General Roberts, the Canadian who commanded the Military Force at Dieppe. We had Gen. Candee talk to us and his speech on the "Air Support in an Attack" is the best I have heard on that subject. We have also had the advantage of having, in this group, Col. Lock, Col. Cleaves and Col. Adams, from Washington, who have been working right in the War Department since the beginning and Britishers who know this work over here. All in all, we have had very fine people. You have something here in this Record which puts your ideas down in concrete form. I assure you that it is going to be used.

The conclusions reached will be of benefit. I want to thank you all for your perseverance. The Assault Training Center is approved. We will have an important mission to perform in our work here. Thank you all very much.

CONCLUDING REMARKS BY: COL. P.W. THOMPSON, COMMANDANT, ASSAULT TRAINING CENTER.

I cannot add a thing to what Gen. Noce has said. Speaking for the Assault Training Center, I want to express my appreciation and the appreciation of all of us in the Assault Training Center, for the good work which has come out of this conference. As Gen. Noce has said, I am sure it is going to be put to good use. It is going to get us off to a flying start, and I think that some day it will "pay off" on the beaches.

APPENDIX

HEADQUARTERS
ASSAULT TRAINING CENTER
ETOUSA (PROV)

14 May 1943.

SUBJECT : Conference on Landing-Assault Doctrine

TO :

1. You are invited to participate in the ETO Conference on Landing-Assault Doctrine, which will begin at 1000 hrs on 19 May with an address by Lieutenant General Jacob L. Devers, Commanding General, ETOUSA. A schedule of meetings for the entire conference and detailed agenda for the first week are enclosed.

2. The mission of the conference will be to develop in detail sound doctrine applicable to landing-assaults on heavily defended shores, with particular reference to cross-channel operations.

3. Several specially qualified officers are coming from the United States to attend the conference and it is considered highly important that well qualified senior officers on duty in this theater participate in it and devote a substantial portion of their time to it while it is in session in order that the doctrine developed may be representative of the best thought available.

For the Commandant:

LUCIUS P. CHASE
Lt. Col, Inf.
Conference Chairman.

HEADQUARTERS
ASSAULT TRAINING CENTER
ETOUSA (PROV)

30 April 1943

INVITATION TO SPEAKERS.

Memorandum to: Speaker's name.

Subject : Conference on Landing-Assault Doctrine for Assault Training Center.

    1. Confirming our conversation regarding the Conference on Landing Assault Doctrine which is being arranged by the Assault Training Center, it is understood that you will address the Conference at ..... hours on ..... May on the subject:"........ in a Landing Assault". It is contemplated that the lecture will take approximately one-and-a-half hours, and will be followed by a conference discussion on the subject which will continue for most of the remainder of the day.

    2. The conference discussion will be guided by an agenda, a tentative copy of which is attached. Your suggestions for its improvement will be appreciated. It would be helpful if this agenda could be returned by 10 May. Although your lecture is not limited to the agenda, and may be much more comprehensive, it should be developed with the agenda in mind.

    3. For your information, the conference discussions will be premised on the following assumptions:

    a. The purpose of the conference is to develop a training doctrine for the Assault Training Center, whose primary mission is to train divisions and their subordinate units for the assault of a heavily fortified Axis-held coast.

    b. The operation contemplated is an invasion, not a raid.

    c. The operation will be cross-channel.

    d. The coastal area attacked will be within fighter aircraft range of airfields in the United Kingdom.

    e. The specific terrain under consideration is the Appledore-Woolacombe training area, fortified to simulate German coast defenses. A sketch map of the area and fortifications will be supplied later.

    4. It is desired that all lectures and discussions include a critical examination of the pertinent parts of FM 31-5, "Landing Operations on Hostile Shores". Frequent reference to the lessons of Dieppe, North Africa and other recent operations will be helpful. In this connection a rather extensive file of reference material is being assembled at Conference Headquarters, Flat 453, 20 Grosvenor Square.

                                     LUCIUS. P. CHASE,
                                     Lt.Col., Inf.
                                     Chief, Int. & Ln.Sec.

Incl: 1 (Agenda)

HEADQUARTERS
ASSAULT TRAINING CENTER
ETOUSA (PROV)

14 May 1943

CONFERENCE ON LANDING ASSAULT DOCTRINE

CONFEREES.

FROM UNITED STATES

Col. Edwin P. LOCK, CE.	Engineer School, Ft. Belvoir
Col. Josiah T. DALBEY, GSC.	Chief of Staff, Airborne Command, AGF.
Col. Haskell H. CLEAVES, SC	Army Section, Amphibious Force, Atlantic Fleet, NOB.
Lt. Col. E. B. GALLANT, GSC	Logistics Group, OPD. War Dept. General Staff.
Lt. Col. Ray ADAMS, GSC	G-3 Division, War Dept. General Staff.
Maj. H. G. SIMMONITE, GSC.	G-4 Sec, OPD, War Dept. General Staff
Lt. Col. A. T. MASON, USMC	COMINCH & CNO, Navy Dept. and MARCORPS
Cmdr. W. H. TURNEY, USN.	COMINCH & CNO, Navy Dept.

FROM EUROPEAN THEATER OF OPERATIONS

HQ, ETO AND SOS

Brig. Gen. D. NOCE, GSC	G-3
Brig. Gen. N. D. COTA	Comb. Opns Liaison
Col. G. B. CONRAD, GSC	G-2 Sec
Col. L. P. HILLSINGER, AC.	Comb Opns Liaison
Col. H. W. GRANT, AC	Comb Opns Liaison
Col. H. V. CANAN, CE.	Eng Service, SOS
Lt. Col. J. B. L. LAWRENCE, AUS	Comb Opns Liaison
Lt. Col. H. E. ZELLER, GSC.	G-2 Sec
Maj. E. H. OSGOOD, FA.	G-5 Sec
Maj. W. L. JAMES, SC.	Sig Service, SOS
Capt. G. B. CAUBLE, SC.	Sig Service, SOS

ASSAULT TRAINING CENTER

Col. P. W. THOMPSON, CE.
Col. W. F. LEE, Inf.
Col. M. W. BREWSTER, FA.
Lt. Col. L. P. CHASE, Inf.
Lt. Col. J. B. HORTON, FA.
Lt. Col. J. T. MARTIN, MC
Maj. A. G. PIXTON, FA
Maj. W. A. BOESMAN, Inf.
Capt. H. J. KELLY, CE.

8TH AIR FORCE

        Brig.Gen.R.C.CANDEE, AC
        Col.Sheffield EDWARDS, AC
        Maj.Wm.McWHORTER, AC.

U.S. NAVY

        Col.W.T.CLEMENT, USMC
        Cmdr.J.S.TRACY, USN

BRITISH

        Maj.M.N.W.BURCH   1 Corps, Home Forces

CONFERENCE ORGANIZATION

NAME	DUTY	FLAT	EXT.
Lt. Col. L. P. CHASE *	Conference Chairman	453F	140
Major J. M. McKEAGUE	Secretary	453G	140
Major W. A. BOESMAN *	Recorder	453G	140
Major A. G. PIXTON *	Recorder	454E	253
Major M. A. PALMER	Supply Officer	453G	140
Tech. Sgt. G. BLACK	Message Center	454B	375
Staff Sgt. GRIFFITH	Librarian	454B	375
Miss K. MASON	Conference Steno	454C	373
Miss I. VIDLER	Conference Steno	454C	373
Mrs. D. MUSCROFT	Conference Steno	454C	373
Miss M. KIRSBERG	Secty: Room No.1	454C	544
Miss J. ROBERTS	Secty: Room No.2	454C	544
Mrs. M. MINNIS	Secty: Room No.3	454C	544

* Lt. Col. CHASE, Major BOESMAN, and Major PIXTON are also members of the conference.

CONFERENCE ON LANDING ASSAULT DOCTRINE.

GENERAL INFORMATION.

<u>Message Center</u>. The Message Center is in the model room (Flat 454B), extension 375. Conferees may use this telephone for incoming calls.

<u>Model Room</u>. Flat 454B has been set aside as a display room, in addition to its use as a Message Center and library. Tech. Sgt. Black will be in charge of this room.

The following will be displayed:

<u>a</u>. Models of old and new landing craft which have been borrowed from the British Combined Ops HQ.

<u>b</u>. Mosaics of beaches on the western coast of France.

<u>c</u>. Pictures of actual German beach defenses, inland and coastal strong points, etc., and two viewing stereoscopes.

<u>d</u>. Terrain model of prospective training area for the Assault Training Center (ETOUSA).

<u>e</u>. Maps and pictures of Fifth Army Invasion Training Center.

<u>f</u>. Other material of interest to the conferees.

TAB "B"

HEADQUARTERS
ASSAULT TRAINING CENTER
ETOUSA (PROV)

21 May 1943

CONFERENCE ON LANDING-ASSAULT DOCTRINE

Date	Time	Activity	Speaker
		PHASE I - ORIENTATION	
May 24 (Mon)	1000	Opening address Orientation talks	Lt Gen Jacob L Devers Col P. W. Thompson Lt Col L. P. Chase
		Films: "Combined Operations" and "Vaagso Raid"	
	1600	G-5 Introduction	Lt Col C. R. Kutz
	1700	Lecture: Combined Operations	Maj Gen J. Charles Haydon Vice Chief, Comb Ops (Br)
	unassigned time	Individual preparation	
		- - -	
		PHASE II - DISCUSSION OF DOCTRINE	
25 (Tues)	0930-1050	Lecture: German coastal defenses and defensive doctrine	Lt Col H. E. Zeller Lt Col Burton (Br) Maj Stamp (Br)
	1100-1230	Discussion	
	1230-1345	Lunch	
	1345-1545	Discussion continued	
	unassigned time	Individual preparation	
		- - -	
26 (Wed)	0930-1050	Lecture: Naval support of a landing-assault	Cmdr. Strauss, USN
	1100-1230	Discussion	
	1230-1345	Lunch	
	1345-1545	Discussion continued	
	1600-1720	Review of previous day's conclusions	
	1730	Lecture: Dieppe Raid Film: "Dieppe"	Commodore Hughes-Hallett, RN
	unassigned time	Individual preparation	

-1-
(Revised 21-5-43)

Date	Time	Activity	Speaker
May 27 (Thurs)	0930-1050	Lecture: Air support of a landing-assault	Brig Gen R.C. Candee
	1100-1230	Discussion	
	1230-1345	Lunch	
	1345-1545	Discussion continued	
	1600-1720	Review of previous day's conclusions	
	unassigned time	Individual preparation	
28 (Fri)	0930-1050	Lecture: Airborne troops in a landing-assault	Col J.T. Dalbey
	1100-1230	Discussion	
	1230-1345	Lunch	
	1345-1545	Discussion continued	
	1600-1720	Review of previous day's conclusions	
	1730	Film: March of Time "We are the Marines"	
	unassigned time	Individual preparation	
29 (Sat)	0930-1050	Lecture: Armored fighting vehicles in a landing-assault	Lt Col C.R. Kutz
	1100-1230	Discussion	
	1230-1345	Lunch	
	1345-1545	Discussion continued	
	1600-1720	Review of previous day's conclusions	
	1730	Lecture: Tanks in a landing-assault	Maj Gen P.C.S. Hobart C.B., D.S.O., O.B.E., M.C., 79th Armoured Division
	unassigned time	Individual preparation	

(Revised 27-5-43)

Date	Time	Activity	Speaker

May
30 (Sun) — Day off

- - -

31 (Mon) 0930-1050 Lecture: Reduction of obstacles and fortifications — Col E.P. Lock

1100-1230 Discussion

1230-1345 Lunch

1345-1545 Discussion continued

1600-1720 Review of previous day's conclusions

unassigned time — Individual preparation

- - -

June
1 (Tues) 0930-1050 Lecture: Artillery in landing-assault — Col G.B. Conrad / Col M.W. Brewster

1100-1230 Discussion

1230-1345 Lunch

1345-1545 Discussion continued

1600-1720 Review of previous day's conclusions

1730 Film: "North African Landings, I"

unassigned time — Individual preparation

- - -

2 (Wed) 0930-1050 Lecture: Infantry in landing-assault — Gen N.D. Cota

1100-1230 Discussion

1230-1345 Lunch

1345-1545 Discussion continued

1600-1720 Review of previous day's conclusions

1730 Film: "North African Landings, II"

unassigned time — Individual preparation

- - -

(Revised 27-5-43)

Date	Time	Activity	Speaker
June			
3 (Thurs)	0930-1030	Lecture: Signal communications for landing-assault	Col Haskell H. Cleaves Col H.W. Grant Maj W.L. James
	1040-1230	Discussion	
	1230-1345	Lunch	
	1345-1445	Lecture: Chemical Warfare	Col H.W. Rowan
	1500-1700	Discussion	
	1715-1815	Review of previous day's conclusions	
	unassigned time	Individual preparation	
4 (Fri)	0930-1030	Lecture: Combined Arms in a landing-assault	Lt Col Ray Adams
	1040-1230	Discussion	
	1230-1345	Lunch	
	1345-1545	Discussion	
	1600-1720	Review of previous day's conclusions	
	unassigned time	Individual preparation	
5 (Sat)	0930-1050	Lecture: Supply and Administration during landing-assaults Film: Amphibious Vehicles"	Maj A.G. Pixton
	1100-1230	Discussion	
	1230-1345	Lunch	
	1345-1545	Discussion continued	
	1600-1720	Review of previous day's conclusions	
	unassigned time	Individual preparation	
6 (Sun)		Day off	

(Revised 27-5-43)

Date	Time	Activity	Speaker
June 7 (Mon)	0945-1000	Remarks on trip to Woolacombe	
	1000-1100	Lecture by Maj-Gen.Hamilton Roberts C-in-Co, Holding Force, Canadian Army, C.O. of Ground Troops at Dieppe.	
	1115-1230	Exercise Kruschen	Brig.O.M.Wales
	1230-1345	Lunch	
	1345-1445	Lecture: Medical service in landing assault.	Col.C.B.Spruit,MC. Col.M.C.Grow, MC.
	1445-1645	Discussion	
	1700-1815	Review of previous day's conclusions	
	evening	Move to Woolacombe	

- - -

PHASE III - PREPARATION OF FIELD EXERCISES

Date	Time	Activity	Speaker
June 8 (Tues) to 13 (Sun)	all day	Committees develop field exercises.	
13 (Sun)	evening	Return to London	

- - -

PHASE IV - ADAPTATION OF FM 31-5 TO ASSAULT TNG CEN'S MISSION

Date	Time	Activity	Speaker
June 15 (Tues)	0930	General Discussion on adaptation of FM 31-5	
	balance of day	Committee work	
16 (Wed)	0930	Coordination Conference	
	balance of day	Committee work	
17 (Thurs)	0930	Coordination Conference	
	balance of day	Committee work	
18 (Fri)	0930	Coordination Conference	
	balance of day	Committee work	

Date	Time	Activity	Speaker
June 19 (Sat)	0930 and all day	Approval of Report on FM 31-5	
20 (Sun)		Day off	

PHASE IV – CONCLUSION

Date	Time	Activity	Speaker
June 21 (Mon)	all day	Committee work	
22 (Tues)	morning	Committee work	
	1400	Talk – "Attu Operation"	Lt.Col.R.O.Bare, USMC
	1530	Conference for approval of notes on FM 31-5	
23 (Wed)	0930	Conference for approval of Training Circular	
	1430	Conference for approval of tentative Conference Proceedings	
	1700	Adjourn.	

For the Commanding Officer:

LUCIUS P. CHASE
Lt.Col., Inf
Conference Chairman.

HEADQUARTERS
ASSAULT TRAINING CENTER
ETOUSA (PROV)

1 June 1943

CONFERENCE ON LANDING ASSAULT DOCTRINE.

Committee Assignments for
Phase III - Preparation of
Field Exercises.

General Committee

Col. Paul W. Thompson, CE.   Commanding Officer
Lt.Col.L.P.Chase, Inf.   Conference Chairman
Maj.J.M.McKeague, FA.   Conference Secretary
Maj.M.A.Palmer   Conference Supply Officer
Lt. D.N.Short, QMC.   Conference Supply Officer
Maj.G.Phillipson-Stow   British Liaison Officer
Col.E.P.Lock, CE.   Engineer Advisor
Col.J.T.Dalbey, GSC.   Airborne Troops Advisor
Lt.Col.J.T.Martin, MC.   Medical Advisor
Lt.Col.H.E.Zeller, GSC.   G-2 Advisor

Committee No.1

Chairman: Brig.Gen.N.D.Cota

Col.W.F.Lee, Inf.
Col.W.T.Clement, USMC.
Col.H.V.Canan, CE.
Col.L.B.Hillsinger, AC.
Lt.Col.E.B.Gallant, GSC.

Committee No.2

Chairman: Col.M.W.Brewster, FA.

Lt.Col.A.T.Mason, USMC.
Maj.H.G.Simmonite, GSC.
Maj.Wm.McWhorter, AC.
Capt.G.B.Cauble, SC.

Committee No.3

Chairman: Lt.Col.J.B.Horton, FA.

Col.S.Edwards, AC.
Col.H.H.Cleaves, SC.
Comdr.J.S.Tracy, USN.
Comdr.W.H.Turney, USN.
Lt.Col.R.Adams, GSC.
Maj.A.G.Pixton, FA.

HEADQUARTERS
ASSAULT TRAINING CENTER
ETOUSA (PROV)

14 June 1943

CONFERENCE ON LANDING ASSAULT DOCTRINE

Directive for Phase IV.
"FM 31-5"

Phase 4 of the Conference on Landing Assault Doctrine will consist of adapting FM 31-5 to the specific task of a cross-channel operation against heavily defended shores. A complete revision of the Field Manual is not contemplated; however, some chapters of it may well be recommended for revision. Other portions may require expansion or amplification. Others may be disregarded as inapplicable.

Committees have been appointed as listed in Appendix "A" attached. It will be noted that several officers are on more than one committee.

Each Committee has been assigned a portion of the Manual for study as indicated below. Major items of responsibility are shown by underlining. Chapters so designated will be given primary attention by the Committee concerned and other chapters listed will be considered of secondary importance. Personnel of Committees which have smaller assignments of work, will be appointed to other Committees upon completion of their initial assignment.

Assignment of Work.

Committee # 1. (General)   Chapters 1, 2 sec.1, 5, & 11, and Assault Technique. (the latter not covered in the Manual, see par.144)

Committee # 2. (Engineer) Chapter 10 sec.VI

Committee # 3. (Signal)   Chapters 5, 8, & 11 sec.III.

Committee # 4. (Naval)   Chapters 2 sec.III, 3, 4, 5, & 6, & appendices.

Committee # 5. (FA & Tanks) Chapters 5, 6, 9, & 11.

Committee # 6. (Air)   Chapters 5, & 7, and Airborne Troops (Col.Dalbey)

Committee # 7. (Supply)   Chapters 2 sec.II, 5, & 10 (less sec.VI)

Committees will work separately, but the full conference will meet daily at 0930 hours for coordination. Coordinating procedure will be as follows:

(a) Each Committee will give the conference secretary, by noon of each day, questions which it needs to have answered by other committees.

(b) The secretary will turn the questions over to the proper committees for study and reply.

(c) Questions will be disposed of directly between the committees concerned, where possible.

(d) Questions requiring discussion or involving several

-1-

committees, or on which a conflict develops, will be placed on the agenda for the coordinating conferences the following morning.

Each Committee will submit a memorandum containing its recommendations on completion of its work. These will be consolidated and submitted to the full Conference for approval.

The schedule will be as shown in the Master schedule for the balance of the Conference. It is subject to adjustment from day to day as work progresses.

HEADQUARTERS
ASSAULT TRAINING CENTER
ETOUSA (PROV)

14 June 1943

CONFERENCE ON ASSAULT-LANDING DOCTRINE

PHASE 4. ADAPTATION OF F.M. 31-5 TO ASSAULT TRAINING CENTER'S MISSION

COMMITTEE ASSIGNMENTS

Committee #1 (General) — (chapters 1, 2 sec. I, 5, and 11 and assault technique)
    Chairman — Col. Lock
    Gen. Cota
    Col. Lee
    Col. Clement
    Col. Dalbey
    Col. Edwards
    Lt. Col. Adams
    Lt. Col. Gallant
    Lt. Col. Zeller
    Maj. Boesman

Committee #2 (Engineer) — (chapter 10 sec VI)
    Chairman — Capt. Murphy
    Col. Lock
    Capt. Kelly

Committee #3 (Signal) — (chapters 5, 8 and 11 sec III)
    Chairman — Col. Cleaves
    Col. Grant
    Maj. James
    Capt. Cauble

Committee #4 (Naval) — (Chapters 2 sec III, 3, 4, 5, and 6) (appendices)
    Chairman — Comdr. Turney
    Comdr. Strauss
    Lt. Col. Mason
    Comdr. Tracy

Committee #5 (FA) — (chapters 5, 6, 9 and 11)
    Chairman — Col. Brewster
    Col. Conrad
    Lt. Col. Horton

Committee #6 (Air) — (chapters 5 and 7)
    Chairman — Col. Edwards
    Col. Hillsinger
    Col. Grant
    Col. Dalbey
    Comdr. Tracy
    Maj. McWhorter

Committee #7 (Supply) — (chapters 2 sec II, 5 and 10)
    Chairman — Lt. Col. Martin
    Maj. Pixton
    Maj. Simmonite

HEADQUARTERS
ASSAULT TRAINING CENTER
ETOUSA (PROV)

16 June 1943

CONFERENCE ON ASSAULT-LANDING DOCTRINE

PHASE 4. ADAPTATION OF F.M.31-5 TO ASSAULT TRAINING CENTER'S MISSION

COMMITTEE ASSIGNMENTS

1. In addition to the committee set up in "Directive for Phase IV", dated 14 June 1943, the following committee is appointed:

    Chairman Col. Lock
         Gen. Cota
         Col. Brewster
         Col. Cleaves
         Col. Dalbey
         Col. Edwards
         Col. Lee
         Lt. Col. Adams
         Lt. Col. Chase
         Lt. Col. Horton
         Lt. Col. Mason, USMC
         Lt. Col. Zeller
         Maj. McWhorter
         Maj. Pixton
         Capt. Murphy

2. It will be called the "Training Circular Committee", and it will take over from Committee #1 the mission of preparing a Training Circular on "Beach Assault Technique".

3. The work of this Committee will be given priority.

4. The Training Circular Committee will meet in the Conference Room

www.ingramcontent.com/pod-product-compliance
Lightning Source LLC
Chambersburg PA
CBHW082114230426
43671CB00015B/2699